WHAT REALLY CAUSES GLOBAL WARMING?

Greenhouse Gases or Ozone Depletion?

PETER LANGDON WARD, PHD

New York

WHAT REALLY CAUSES GLOBAL WARMING?
Greenhouse Gases or Ozone Depletion?

© 2016 PETER LANGDON WARD, PHD.

Published in New York, New York, by Morgan James Publishing. Morgan James and The Entrepreneurial Publisher are trademarks of Morgan James, LLC.
www.MorganJamesPublishing.com

The Morgan James Speakers Group can bring authors to your live event. For more information or to book an event visit The Morgan James Speakers Group at
www.TheMorganJamesSpeakersGroup.com.

Cover photo: The volcano Bárðarbunga erupting in central Iceland on September 4, 2014. From August 29, 2014, through February 28, 2015, this volcano extruded the Holuhraun lava field, covering an area of 33 square miles (85 km²), the largest basaltic lava field observed in the world since the Laki eruption in 1783. This volcanic eruption appears to have had profound effects on weather, including making 2014 and 2015 the warmest years on record. Photo © Arctic-Images/Corbis

Shelfie

A **free** eBook edition is available
with the purchase of this print book.

CLEARLY PRINT YOUR NAME ABOVE IN UPPER CASE

Instructions to claim your free eBook edition:
1. Download the Shelfie app for Android or iOS
2. Write your name in **UPPER CASE** above
3. Use the Shelfie app to submit a photo
4. Download your eBook to any device

ISBN 978-1-63047-798-1 paperback
ISBN 978-1-63047-799-8 eBook
ISBN 978-1-63047-800-1 case laminate
Library of Congress Control Number:
2015915175

In an effort to support local communities and raise awareness and funds, Morgan James Publishing donates a percentage of all book sales for the life of each book to Habitat for Humanity Peninsula and Greater Williamsburg.

Get involved today, visit
www.MorganJamesBuilds.com

Habitat
for Humanity®
Peninsula and
Greater Williamsburg
Building Partner

WHAT REALLY CAUSES GLOBAL WARMING?

TABLE OF CONTENTS

TABLE OF FIGURES

Chapter 4: Do We Really Understand Thermal Energy?

Chapter 5: How Does the Atmosphere Protect Earth From Sun's "Hottest" Radiation?

Chapter 6: How Do Minute Amounts of Ozone Control Climate?

Chapter 7: How Does Temperature Change With Ozone Depletion?

Chapter 8: How Can Volcanoes Both Cool and Warm Earth?

Chapter 9: How Do Volcanic Eruptions Affect Weather?

Chapter 10: Why Does the Greenhouse Effect Appear Not to be Correct?

Chapter 11: What Are Some Other Implications of Light Being a Continuum of Frequency?

Chapter 12: How Could Science Have Been So Far Off the Mark?

Chapter 13: Where Do We Go From Here?

A MORE PERSUASIVE TRUTH

David Bennett Laing
Assistant Professor of Geology, retired, University of Maine
Author: *The Earth System: An Introduction to Earth Science*

"The free, unhampered exchange of ideas and scientific conclusions is necessary for the sound development of science, as it is in all spheres of cultural life."
—**Albert Einstein**, 1952

In the fall of 2014, while waiting to view a film about ocean acidification with a conservation group in Belfast, Maine, I got into a conversation with a young member of the group about whether or not the science of climate change really is "settled." I admitted to her that I was somewhat skeptical. At issue in particular was the so-called "hiatus" in global warming—the enigmatic, seventeen-year period since 1998 during which the increase in global warming seems to have either stopped or slowed markedly, despite the ongoing dramatic increase in emissions of **carbon dioxide** into Earth's atmosphere. She seemed quite happy to debate the question with me in a congenial manner, but as the discussion went on, I noticed that her much older husband, who was sitting next to her, was becoming increasingly agitated and uncomfortable. Finally, he could take it no longer and erupted with a volley of vitriolic language, accusing me of wrong-headed thinking bordering on sociopathic behavior. I hastened to end the conversation and to find myself a seat for the impending movie, but as I turned away, I was cornered by another, younger man who had been standing by, listening in. He carried on in much the same vein, painting me as a shill for Big Oil and an enemy of the people.

To me, a dedicated populist activist, and one who often signs online petitions to curb the excesses of Big Oil and other Wall Street operatives, this came as a bit of a shock. In the course of my research and teaching activities as an Earth systems scientist, I had come across discussions in the literature now and then that questioned the validity of greenhouse warming theory, and I had long assumed that it was a valid topic for debate. Now, however, I was coming up against the hard reality that it had become a highly contentious, hot-button issue. It was a real eye-opener for me, and from that day forward, I began paying more attention to the human dimension of the greenhouse warming debate and was quite surprised to discover the extent to which it had become both polarized and politicized.

More than anything else, however, what this incident did for me was to convince me that **anthropogenic** (human-caused) climate change is one of the most important scientific and public policy issues of our times, potentially affecting all life on Earth. That conviction only heightened my concern over the problematics of greenhouse warming theory. Clearly, it is of the utmost importance that we get this one right. Are we really on the right track? If not, what needs to be done in order to get us there, and once there, what should we do about the problem?

I had become generally aware of the growing polarization over global warming, and the rising political stakes, through my progressive online activism. Among the many different petitions on which I took action, there were some that called for reductions in **greenhouse gas** emissions. These had always given me pause because of my long-held awareness of issues in greenhouse warming theory, but I usually signed them anyway, until the fall of 2012, when I received an email from my old Dartmouth College friend and geological colleague Peter Ward. He had attached a new paper he had written proposing the novel idea that instead of carbon dioxide, a far more likely driver of global warming was chlorine from **chlorofluorocarbons** (CFCs), which were released into the atmosphere during the last three decades of the 20th century, an interval in which the observed planetary warming was far more dramatic than it was before or has been since. The chlorine wound up destroying ozone in the **stratosphere** until CFC production was halted by the **Montreal Protocol** on Substances that Deplete the Ozone Layer, which went into effect in 1989. In the paper, Peter suggested that the thinned **ozone layer** let in an excess of solar ultraviolet-B (UV-B)

radiation, and that this excess of that high-energy radiation should easily account for the observed global warming.

This conceptual model, he explained, had germinated from his extensive studies of global volcanism in his capacity as a geophysicist with the United States Geological Survey. In the course of his work, he had noticed that prolonged periods of frequent, intense **basaltic** volcanism were consistently associated with episodes of pronounced global warming, over recent geologic time, and combining this with other observations and reasoning, he concluded that the release of chlorine during such eruptions depleted stratospheric ozone, allowing increased input of solar **ultraviolet radiation**, which produced global warming.

This was breakthrough science. It was already well established that **explosive**, andesitic **volcanoes** hurl water vapor and **sulfur dioxide** into the stratosphere, forming **aerosols** that block sunlight and cause global cooling, but no one had ever proposed that there was a warming effect from the chlorine that both andesitic and basaltic volcanoes emit. In the case of explosive volcanoes, Peter explained, that warming effect was overwhelmed by the cooling effect of aerosols, resulting in net cooling. Using very detailed graphics, he was able to use this elegant conceptual model to explain all the enigmatic warming and cooling events of the past 100,000 years of Earth's history, something over which Earth scientists have debated contentiously for decades.

Skeptical at first, I re-read the paper with a view to finding fault with it, but quickly realized that I couldn't. The fundamental argument and all its supporting data were not only internally consistent but also fully consistent with all the pertinent facts (as opposed to theories) of climate science with which I was familiar. Peter's argument, I realized, offered a far more rational and compelling explanation for the phenomena of global warming and cooling than did greenhouse warming theory. That marked the end of my signing petitions calling for draconian restrictions of carbon dioxide emissions. Why, I reasoned, should the world enter into a massive and expensive global campaign to curb greenhouse gas emissions if carbon dioxide and other greenhouse gases (chiefly water and methane), aren't actually what drive global warming? That would be about as effective as trying to extinguish an electrical fire with water instead of just turning off the electricity. Being the wrong solution, it wouldn't solve the problem, and it would likely make things a whole lot worse.

I had no moral or political agenda in making this switch. My principal motive was, and remains, a strong commitment to seeking the most accurate possible interpretation of reality and truth. A secondary motive was, and remains, an aversion to counterproductive policies made on the basis of flawed interpretations. I realized that Peter's conceptual model does a better job of explaining the observed facts than does greenhouse warming theory, and that, as far as I could tell, it raises no issues with fundamental physical laws, as I have long felt greenhouse warming theory does. On the other hand, I fully recognized that introducing Peter's variant view in what is clearly a highly polarized and politicized arena would be challenging, to say the least. I also recognized, however, that Peter's hypothesis seems to lie in a middle ground between the two camps, a strategic position from which he might be able to reconcile some or perhaps even all of the differences that have led to their extreme polarization.

What are these two camps? On the one hand, there is the mainstream academic climate science community, which is firmly committed to greenhouse warming theory and to the concept that a continued increase in the atmospheric concentration of anthropogenic carbon dioxide has led to, and will continue to produce, a corresponding increase in global temperature. Peripheral to and supporting this camp is a large and dedicated group of mainly progressive, non-scientist activists, increasingly joined by academics, including even some climate scientists, who view the specter of greenhouse warming with a degree of alarm that often approaches religious intensity. One catchphrase that is often associated with this camp is "settled science," a concept that should be anathema to any scientist worth his salt, as it flies in the face of the principle that in properly conducted science, nothing is ever "settled." Despite this, I soon realized that many climate scientists are actually making that very claim. The opposition has, with some justification, adopted the phrase "settled science" as a derisive term.

The other camp, which has been styled "climate deniers" by the more politically motivated element in the first camp, consists largely of conservatives, champions of unfettered free enterprise, and stakeholders in industries that produce or consume fossil fuels. These include a few climate scientists, who maintain that any variation in global **temperature** over time is due not to human activity but to natural climate variability. Coupled with this view is the conviction that the "settled scientists" have cooked the books, cherry-picking and even altering, or at least statistically manipulating data in order to tweak the historical climate record in ways that tend to support their contention that anthropogenic climate change is real and has overprinted and overwhelmed the effects of natural climate variability.

Peter Ward's position lies, as I have suggested, somewhere in between these two extremes. It acknowledges the likely reality of anthropogenic global warming and consequent climate change, a view that is more in line with the "settled science" camp, but it questions the validity of greenhouse warming theory, which is more in line with the "climate denier" camp. It would be premature, perhaps, to think that Peter's middle-ground stance could, of itself, bring about a rapprochement between the two camps, but his stance is strong enough that I feel there is some reason for optimism.

In the months between my first reading of Peter's paper and the incident at the Belfast Library, Peter sent me a series of revisions and ultimately the link to his new website, ozonedepletiontheory.info, which explains his theory not only in terms of the geological evidence, but also presents some remarkable insights into the nature of **electromagnetic radiation** that are central to the question. At first, I had trouble understanding the relevance and the significance of these insights, and I felt that they might actually detract from the credibility of his main argument. I even went so far as to urge him to drop the physics and to concentrate instead on his highly compelling geological argument, but he was adamant. Frustrated, I finally suggested that maybe I could contribute to his campaign in an editorial capacity, in which I had considerable professional experience. In particular, I suggested that we could write a popular book to showcase his ideas, which he had so far been unable to publish in a peer-reviewed scientific journal because of their contradiction of the universally accepted greenhouse warming theory. He accepted, and in the course of working with him, I was eventually able to grasp the meaning and the import of the physical insights, and their relevance. After passing that hurdle, I put a special effort into clarifying the discussion of the concepts for ease of comprehension.

That said, although I may have contributed to the refinement and clarity of concepts throughout our collaboration, this is entirely Peter's book, and it is, in my opinion, a brilliantly conceived tour-de-force of science that should not only help to rectify a serious misconception in climate science, but also to stimulate serious new thought in the fields of radiation physics and **quantum mechanics**.

PREFACE

"The first guessed at nature rather than studying it; the others, while thinking they are only verifying the systems they admire, study it truly; and it is thus that the sciences—like peoples—pass from poetry to history."
—Georges Cuvier, 1800

In 2006, while enjoying retirement, rafting, climbing, skiing, and folk-singing in Jackson, Wyoming, I came across an enigma, a puzzling thing about climate change that just did not make sense. Based on my lifelong involvement with **volcanoes**, earthquakes, Earth science, geophysics, physics, public policy, public education about science, and yes, also some paleoclimatology and meteorology, I had a gut feeling that resolving this enigma would not only be important, but might provide improved understanding of climate and weather—past, present, and future. I had no idea then where I would end up—I was just curious about something that I was convinced could become important.

After carefully looking into the details, I decided to put almost everything else aside in my life, except for my family, to try to understand this enigma. As I dug deeper and deeper, I found more puzzles that just didn't add up. I began to look systematically at a wide variety of assumptions underlying the greenhouse warming theory of climate change. I began to look at assumptions underlying the basic physics of **radiation**—the physics of light. These were all assumptions that were accepted as most likely true by most scientists involved. As an Earth scientist, as a geophysicist, and as someone who has done a great deal of field work around the world, I am a stickler for observations. I want to see them, sense them, feel them, either personally, using instruments, or by reading about observations made by others that appear to be robust. I view statistics and mathematics as tools, not gospel. I look at theory as possibility, not truth. What started as curiosity developed into an adventure, a search, an odyssey, and ultimately a quest.

In this book, I invite you to join me in retracing some key paths and terrains on this quest, looking at data collected meticulously by thousands of scientists, asking how things work, asking what could be the cause of the observed data. I became very passionate about getting to the bottom of some key issues. I hope you will too. My greatest joys in life are

Peter Ward, second from the left, leading the field work in barren, remote, central Iceland in 1968 that formed the major part of his PhD thesis. The location is about ten miles west of Bárðarbunga, the major effusive volcano that erupted in 2014. On the left, Lynn Sykes, who showed in 1966 that earthquake mechanisms agreed with Tuzo Wilson's hypothesis of transform faults and ultimately became a professor at Columbia University. On the right, looking where we were headed, Páll Einarson, my key Icelandic field assistant, now Professor at the University of Iceland. To his left in front, John Kelleher, a fellow graduate student from Lamont-Doherty Geological Observatory at Columbia University. Photo by Sveinn Magnússon, another Icelandic field assistant now Director General at the Icelandic Ministry of Welfare.

typically when I get deeply involved in something that I think is important and find myself on a steep learning curve. As Albert Einstein is reputed to have said, "intellectual growth should commence at birth and cease only at death."

I will share with you graphs and plots of a large amount of scientific data, but you do not need to have training in science. You only need to be curious and to roll up your sleeves to do the work—to share in the fun of discovery. Gaining new insights into problems we ponder can be a lot of fun. Science is a lot of fun, too, even though, like most worthwhile things in life, it does take hard work.

One key result of my work, explained in this book, is that the greenhouse warming theory of climate change does not appear to be correct. That assessment is based on questions about some critical assumptions made about light (**electromagnetic radiation**) in the 1860s, more than 150 years ago, that I feel are of questionable validity. Understanding what light actually is and how it actually radiates has major implications not only for climate change, but also for modern physics.

I hope to convince you in this book that ozone depletion explains most observations of climate change quite well and that it is physically impossible for the **infrared** energy that is radiated by Earth and absorbed by greenhouse gases to play a significant role in global warming. Certainly, there are many who would view such claims as very inconvenient truths!

If I am right, these conclusions could very well be the worst nightmare for those climate scientists who have dedicated their lives, working long and hard with a strong sense of public service, to convince the world to take action before we destroy our livable Earth. The vast majority of climate scientists are convinced beyond any reasonable doubt that global warming is fully explained by greenhouse warming theory. These ideas have been developed theoretically over the past 150 years, and I have reviewed that distinguished history in considerable detail. There is a lot of very good science that has been built up meticulously on the foundation that existed when all scientists now living came of age. Unfortunately, however, that foundation is not perfect. While scientists seek perfection, they are, ultimately, human. Sometimes, new data and insights require the rethinking of old conclusions. What most good scientists get passionate about is seeking new and improved understanding. My sincere hope is to be able to share with other scientists my own passion in making these new discoveries.

Especially over the last 20 years, climate scientists have been under increasingly vitriolic attack by people they view as "skeptics," ranging from other scientists who believe our understanding of climate science may be incomplete to those who categorically deny that climate change is occurring. In any debate in life that threatens the status quo, and especially that threatens our healthy future, participants muster whatever arguments they can. Some roll out the cannons, others talk calmly and rationally, putting their spin on the data. Some, for lack of anything constructive to say, launch personal attacks. Others declare strength in numbers. This is, after all, the way human politics works, but is it the way science should work?

The last thing scientists want, with wagons circled, fending off attacks, is for one of their own suddenly to raise a red flag and tell them that unfortunately there is a problem with their science. Scientists know as well as anyone that science seeks perfection but rarely gets it on the first try. Science is a way of collecting, organizing, analyzing, and interpreting data in a logical manner with hopes of understanding how things in Nature, and in human nature, work. What are the causes and what are the effects? Data are seldom complete, and they are seldom completely reliable. We observe results, but we have to infer causes, using the best logic we can. Interpretations typically are built into robust conclusions, but sometimes we diverge on a detour that we might not recognize as such for a long time.

Often, in science, it is an outsider who sees the obvious—the generalist who is not imbued with all the traditions of a specific field. We all craft our own world views, through our own learning, through our own research, and through interactions with others. We all build our own foundations upon which to build our futures—upon which we can make decisions about how to think, how to behave, and when to take action. Without such working hypotheses, we might be paralyzed with too much indecision to move forward. We also defend our working hypotheses. We have to develop spam filters—ways of sorting the wheat from the chaff. Information is traveling so fast these days that we are becoming victims of our spam filters. There is just too much information available to think about and only so much time to think.

The United Nations Intergovernmental Panel on Climate Change (IPCC) was founded in 1988, almost three decades ago, because many scientists were convinced that the world is warming, that much of the warming is caused by man, and that politicians need to take action now before the world becomes dangerously hot. Members of the IPCC have worked hard to demonstrate that there is consensus among a large number of scientists, and that the science is therefore "settled." That, however, is a highly unscientific conclusion. Science is never settled. There were times when most scientists agreed that the world was flat, that Sun revolved around Earth, that Earth was 4004 years old, that continents were fixed and immobile, and that life forms were fixed and do not evolve. These all seemed to be reasonable theories in their time, based on clear observations. New observations, new concepts, and new ways of thinking, however, have often led to new theories that modern scientists think are closer to reality. If everything were settled, there would be no need for science. Charles Duell, Commissioner of the US patent office in 1899, is reported to have said "everything that can be invented has been invented." Imagine where we would be today if all aspiring inventors had believed him.

While politics requires consensus, science thrives on debate. If climate science is settled, then why continue to pay climate scientists just to dot the "i"s and cross the "t"s? The goal of science is to acquire better understanding of how things actually work and, in the process, to help societies and individuals adapt to changes in the world around us.

Alfred Wegener was a meteorologist, geophysicist, and polar researcher who hypothesized in 1912 that the continents were slowly drifting around the world. As maps of the world became more accurate, many people began to realize that if you were to slide South America eastward toward Africa, the continents would fit together like a good jigsaw puzzle, but Earth scientists "knew" that continents are fixed in position—that continental drift is impossible. They gave Wegener endless grief over his hypothesis. Wegener died in 1930, and it was not until the 1960s, when huge amounts of scientific data about ocean floors were collected, that most Earth scientists finally came to accept the reality of seafloor spreading, the process by which continents drift around Earth's surface.

Many friends have pointed out to me that maybe it is only fair that meteorologists and climatologists give me, an Earth scientist, the same kind of grief that Earth scientists gave the meteorologist Wegener. Nevertheless, I do hope that with the wealth of data currently available, they might see some value in my ideas before I die.

So welcome to science! Thank you for having the interest to learn from observations, the closest thing to ground truth in science. Let's get on with the adventure.

Notes for the Reader

Most technical words used several times throughout the text are shown in **bold font** when they are first used in each chapter and are defined in the Glossary at the end of the book. Sometimes, I follow a word with a different but similar word in parentheses to help emphasize or clarify a point.

There are also many endnotes in each chapter to help you access more information on specific topics, if desired. I have tried, in general, to choose references that would be most useful to nonspecialists. There are many more scientific references attached to each of these topics on my website ozonedepletiontheory.info.

I give many references to Wikipedia, which some people think of as unprofessional because anyone can write anything that they want in Wikipedia. The open-source software movement, out of which Wikipedia has grown, has revolutionized our lives, saving consumers $60 billion per year with such widely utilized programs as Firefox and Thunderbird and operating systems such as Linux.[1] What makes open-source work, is that most humans want to be appreciated and respected by their peers. They want to do a professional job. People who take the time and effort to contribute to Wikipedia, typically do an amazingly good job provided they have no hidden agenda, and the Wikipedia procedures and staff try to keep such problems to a minimum. Wikipedia has a peer group that discusses controversial pages in great detail and editors that encourage appropriate improvements. Wikipedia competes quite well with Encyclopedia Britannica for accuracy[2] and outranks any printed encyclopedia substantially for numbers of quality articles. I find that, in science at least, Wikipedia articles are generally done quite well and that the introduction of each Wikipedia article typically answers initial questions, while more than enough detail follows for those interested. Furthermore, all Wikipedia articles are extensively cross-referenced with hyperlinks.

I have included many references to published papers. They include a Digital Object Identifier (DOI) code assigned by each publisher. If you go in your browser to dx.doi.org/ and type in the code, you will be taken to that specific paper on the Internet.

Most scientists and most people in the world use the metric system, but I have used units most typical in America and have included the metric equivalents in parentheses.

Much more detail about everything in this book is provided at ozonedepletiontheory. info, with many more references dynamically linked through the DOI system and with hyperlinks to definitions of many of the technical terms. This website was written primarily for scientists, but non-scientists may also find it valuable.

This book is described with links to other material on its website WhatReallyCausesGlobalWarming.com. You can find out more about me at PeterLWardPhD.com. All of these sites can link you to a number of short videos, interviews, longer talks, and documentaries concerning ozone depletion.

Serious feedback on this book would be welcome. Please send any scientific comments or questions to info@OzoneDepletionTheory.info.

Special Thanks

I met David Laing in 1961, during my first month at Dartmouth College. He and his wife were singing folk songs to a large audience in the largest auditorium then available in Hanover. He was three years ahead of me. I learned from him a lot of folk songs, rock climbing, and skiing. We did some winter mountaineering trips together. We reconnected a few times over the decades, but since he responded to my draft paper in 2012, we have developed a working relationship that not only led to this book, but has improved its quality immeasurably. David has questioned all of my ideas, providing a counterpoint required for good science. He has pushed me to explain my ideas more clearly and he is an outstanding copy editor. While we discussed co-authorship, we decided that the book needed to be my personal story. I sincerely appreciate our friendship and all of his hard work.

Mike McCracken and Peter Molnar have provided very detailed reviews of the many scientific papers that I have written over the past 9 years. They worked hard to hold my feet to the fire, refining my science. As I began to question greenhouse warming theory, their interest began to wane, but their critical thinking over many years has been invaluable. Thank you both very, very much.

I sincerely appreciate the endless reviews and edits provided daily by my wife, Adrienne Ward, and for her gracious understanding of why I spend so much time at my desk.

Detailed reviews of this book have been provided by Doug Ayers, Erhard Bieber, James Bjorken, JoAnn Blomberg, Jane Brown, Roger Brown, John Good, Jim Herriot, Wes Hildreth, Phil Hocker, Ernie LaBelle, Lillian McMath, Karen Reinhart, Dan Shyti, Tom Smith, Bob Tilling, Steven Unfried, Barbara Waters, and John Willott. Comments, both agreeing and disagreeing, have been received from Joe Burke, Ed Henze, Bruce Julian, David Mittell, Bert Raynes, and Warren Hamilton.

Thanks to John Sharsmith for meticulously compiling distribution information.

Figure 8.21 Figure 8.21 was created by Steven J. Epstein, www.sjepsteindesign.com. Thank you, Steve.

All of these people have helped improve this book, but I take personal responsibility for the concepts and for the words.

Peter Ward, February, 2015

CLIMATE IS NEVER "SETTLED," AND NEITHER IS SCIENCE

"The work of science has nothing whatever to do with consensus. Consensus is the business of politics. Science, on the contrary, requires only one investigator who happens to be right, which means that he or she has results that are verifiable by reference to the real world."

—Michael Crichton, 2003

Climate is always changing, usually so slowly that humans can readily adapt, but sometimes so quickly that many people die and civilizations fail for lack of sufficient resources such as water, food, and healthy air. Our species, *Homo sapiens*, is the child of climate change. At least twenty-five times from 100,000 to 10,000 years ago, every 3600 years on average, the world warmed rapidly by 18 to 29°F (10 to 16°C) out of an ice age. Within a decade or two—only part of a human lifetime—temperatures escalated many degrees, glaciers melted, and sea level rose, followed by a slow drift back into ice-age conditions over a century or more—over several human lifetimes. Our ancestors proved resourceful. Apparently, they were better able than the Neanderthals and other human-like species to adapt their diet, to migrate, and to learn to work together in order to survive these sudden changes. As more and more data are collected, we are finding a stronger and stronger link between climate change and change among living things throughout geologic history, throughout the evolution of our species, and throughout recorded history. Our ancestors, who were the ones that survived, obviously figured out how to adapt.

Recently, we have been hearing news stories almost daily about global warming or about record-setting severe weather. What is going on and why? Are humans to blame? Will it get worse? What can we do about it? What should we do about it? Finally, there is the very personal question: how should I and my family adapt?

This book starts off by examining the rapid global warming that clearly did occur between 1970 and 1998 and that, in my view, appears to have been caused by humans depleting the **ozone layer**, located 12 to 19 miles (20 to 30 km) above Earth's surface. In 1995, Mario J. Molina, Paul J. Crutzen, and F. Sherwood Rowland received the Nobel Prize in Chemistry for their "work in atmospheric chemistry, particularly concerning the formation and decomposition of ozone."[3] Rowland, in his acceptance lecture, said "the ozone layer acts as an atmospheric shield, which protects life on Earth against harmful **ultraviolet radiation** coming from the sun. This shield is fragile: in the past two decades it has become very clear that it can be affected by human activities."[4] When the ozone layer is thinned (depleted), more ultraviolet radiation than usual reaches Earth, raising surface temperatures.

Global warming by one degree Fahrenheit (0.6°C), between 1965 and 1998, appears to have been caused by a rapid increase in manufacturing of very stable **chlorofluorocarbon** gases (CFCs) used in aerosol cans, refrigerators, air conditioners, and certain fire extinguishers and used as solvents and foam blowing agents. Mario Molina discovered that when CFCs rise up into the coldest environments of the lower **stratosphere**, they are broken down by highly energetic, solar ultraviolet radiation, releasing chlorine that depletes ozone, causing the **Antarctic ozone hole**, major ozone depletion in the Arctic, and more moderate ozone depletion at mid-latitudes.

When the Antarctic ozone hole was discovered in 1985, scientists and political leaders worked well and promptly together to pass the **Montreal Protocol** on Substances that Deplete the Ozone Layer. This protocol became effective in 1989, ending the increase in emissions of CFCs by 1993, ending the increase in ozone depletion by 1995, and, as I argue in this book, ending the increase in mean surface air **temperature** by 1998. Humans inadvertently caused ozone depletion, and humans acted knowledgeably and decisively to end the increase in ozone depletion, which also happened to end the increase in global warming. If the Montreal Protocol had not taken effect, global mean surface air temperatures would likely have continued to rise after 1998.

Chlorine and bromine gases emitted by all types of **volcanoes** also deplete the ozone layer. **Effusive** volcanoes pour vast quantities of black, **basaltic** lava out onto Earth's surface, as is typical in Hawaii and Iceland, releasing chlorine and bromine that deplete ozone and cause global warming. **Explosive** volcanic eruptions, on the other hand, such as Mt. Pinatubo in the Philippines in 1991, also deplete ozone, causing winter warming, but in addition, they explode megatons of gases into the stratosphere, just below the ozone layer. These gases, especially water vapor and **sulfur dioxide**, form aerosols, a fine mist, whose particle sizes grow large enough, over several months, to reflect and scatter sunlight for a few years, overwhelming the warming effect and thus causing net global cooling.

Large, explosive volcanic eruptions currently occur once or twice per century. When they happen more frequently—many times per century—they can, over a few thousand years, increment the world into an ice age. Conversely, extensive, voluminous, effusive, basaltic volcanism can warm the continents out of an ice age within a few years and can warm the oceans—Earth's principal heat reservoir— out of an ice age within a few thousand years, thus explaining the climate oscillations faced by our ancestors.

Ozone is expected to remain depleted for many more decades, causing the oceans to continue to warm. The longer ozone remains depleted, the warmer the oceans will become, and the longer Earth's warmer climate will persist. In fact, there is no natural way to cool climate back to pre-1975 levels except as a result of an increased frequency of large, explosive volcanic eruptions.

Ozone concentrations vary rapidly, especially in late winter and early spring, having major effects on weather, on the basic distribution of air-pressure highs and lows, and on the location and strength of each **polar vortex** and of jet streams.

The greenhouse warming theory of global warming, which most scientists consider "settled," i.e. proven beyond any reasonable doubt, is not only mistaken—it is physically impossible. Why? In the first place, greenhouse warming theory says that infrared radiation from Earth warms the planet more than ultraviolet energy from Sun. Common experience confirms that this is not true—you get hotter standing in sunlight than outside at night with Earth's infrared radiation welling up around you. Secondly, thermal energy in a colder atmosphere cannot raise the temperature of a warmer planet—you do not get warmer standing next to a cold stove. Third, radiation from a thermal body cannot warm the same body—otherwise, all material objects would spontaneously heat up, which they obviously do not.

Greenhouse warming theory builds on observations by John Tyndall[5] in 1859 that **carbon dioxide** gas absorbs infrared thermal energy radiated by Earth. It seems logical that increased concentrations of carbon dioxide in the atmosphere would increase the thickness of a "greenhouse gas blanket" surrounding Earth, thereby keeping Earth warm. Surprisingly, however, there are reasons to question this assumption, as I will discuss in Chapter 4. It has never been proven in the laboratory that increasing the concentration of any **greenhouse gas** in air actually increases the temperature of the air as much as observed.

Many scientists have pointed out that Earth's average surface temperature is approximately 60°F (33°C) warmer than a planet should be that is located 93 million miles (150 million kilometers) from Sun. Thus, they say, Earth's atmosphere must be keeping Earth warm. Well, it does that, but not because of greenhouse gases. The stratosphere forms an "electric blanket" in the sense that the energy causing the observed increase in temperature with altitude in the stratosphere comes from elsewhere, from Sun, not from the blanketed body, Earth.

The stratosphere is warmed by high-energy solar ultraviolet radiation that continually dissociates (splits apart) molecules of oxygen (O_2) to form atoms of oxygen (O), releasing heat. Nearly all of this high-energy ultraviolet-C radiation is absorbed and converted to heat in the stratosphere. A molecule of oxygen (O_2) and an atom (O) of oxygen then combine to form a molecule of ozone (O_3). Slightly lower energy ultraviolet-B solar radiation then splits ozone apart into an atom and a molecule of oxygen, also releasing heat. The average lifetime of one molecule of ozone is only 8.3 days. This conversion back and forth goes on continually on the sunlit side of Earth, turning solar ultraviolet radiation into heat in the stratosphere. Wherever ozone exists, heating of the atmosphere is going on.

Earth is kept warm because all of the atmosphere above the **tropopause**, 5.6 mi (9 km) above the poles to 11 mi (18 km) above the tropics, absorbs very high energy, very high-frequency, very "hot" ultraviolet radiation from Sun, protecting life on Earth from this DNA-destroying radiation. Radiation does not have temperature, but it carries the potential to heat matter; the higher the frequency of radiation, the higher the temperature becomes in the absorbing matter. When the ozone layer just above the tropopause is depleted (thinned), more high-energy, high-frequency, ultraviolet-B radiation is observed to reach Earth, warming Earth and increasing your risk of sunburn and skin cancer.

In addition, volcanoes are adding to our current problems. The volcano Bárðarbunga in central Iceland, featured on the cover of this book,[6] erupted 33 square miles (85 km²)[7] of black, basaltic **magma** from August 2014 to February 2015, the largest effusive volcanic eruption in the world since the ten-times more voluminous eruption of Laki in 1783 that killed tens of thousands of people in Europe, possibly triggering the French revolution.[8]

The eruption of Bárðarbunga is the most likely cause of recent increases in average global surface temperatures.

Let's now look at the evidence. Are these conclusions "verifiable by reference to the real world"[9]?

CHAPTER 1

HOW I CAME TO WONDER ABOUT CLIMATE CHANGE

"In the fields of observation, chance favors only the prepared mind."
—Louis Pasteur, 1854

What should I do if this **volcano** starts to erupt? I'm 19 years old, climbing to the summit of an active volcano in Alaska with my Dartmouth College professor and life-long mentor, Bob Decker. He was already recognized as a world-class expert on active volcanoes, and he went on to be Scientist in Charge at the Hawaiian Volcano Observatory and President of the International Association of Volcanology and Chemistry of the Earth's Interior.

We are walking up the side of Mt. Trident, a volcano named for its three prominent peaks, but we are on a new, fourth peak that has grown nearly 1000 feet (300 meters) vertically in the past ten years. We've studied this volcano from distant hill-tops, from circling around it in a very small airplane, and from pictures of recent eruptions in the files of the National Park Service. Now we are on the ground, exploring the details up close and personal, walking where very few people have been before.

Katmai National Monument, 250 miles southwest of Anchorage, Alaska, is the size of the state of Connecticut, but it is remote country, with only one 23-mile-long dirt road, accessed by float plane. This one-lane, primitive road, still under construction in 1963, heads southeast from Brooks Camp, taking tourists to an overlook from which they can see the Valley of 10,000 Smokes and many volcanoes. We wave goodbye, to the bus driver, shoulder our packs, and hike south for two days through the Valley of 10,000 Smokes and through Katmai Pass for our first glimpse of Trident volcano (Figure 1.1).

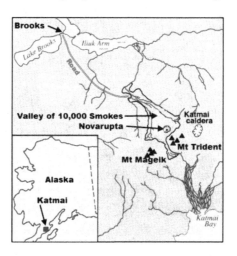

Figure 1.1 In 1912, Mt. Katmai erupted, disgorging a huge pyroclastic flow from Novarupta, forming the Valley of 10,000 Smokes. Mt. Trident was just to the south. The red line shows the road from Brooks Camp to the Overlook.

The Valley of 10,000 Smokes was named by Robert Griggs, leader of a National Geographic Society expedition in 1916, whose objective was to explore the source of the 1912 eruption of Mt. Katmai, the largest volcanic eruption of the 20th century. Griggs had landed by ship in Katmai Bay, and had then hiked northward along the Katmai River Delta, up past Mt. Trident to Katmai Pass. "The sight that flashed into view as we surmounted the hillock was one of the most amazing visions ever beheld by mortal eye," he wrote. "The whole valley as far as the eye could reach was full of hundreds, no thousands—literally, tens of thousands—of smokes curling up from its fissured floor."[10] A frothy cloud of volcanic gases and rock fragments, variously referred to as a **pyroclastic** flow, a glowing avalanche, or a nuée ardente, had flowed out of the Novarupta vent, filling the valley within minutes with deposits that, when cooled, would become a dense, flinty rock called **welded tuff** capped with a light, frothy rock called **pumice**. In 1965, I measured a thickness of the valley fill in one prominent location as 98 feet (30 m), after substantial erosion, but some think that the pyroclastic flow could have been as much as 689 feet (210 m) thick in places.[11] A similar flow in 1902, just ten years before the Katmai eruption, had buried the town of St. Pierre in the Lesser Antilles arc of the Caribbean, killing 30,000 people instantly. In 1963, the Valley of 10,000 Smokes had just a few steaming vents left, and they were barely hot enough to form steam.

In 1953, Trident volcano began forming a new volcanic cone on its southwest flank. The cone had grown 853 feet (260 m) by 1960. Some **ash** eruptions had reached altitudes of more than 5.6 mi (9 km). Viscous, blocky black lava flowed out of the volcano's vent throughout the 1950s forming flows as thick as 984 feet (300 m), covering 2 mi^2 (5 km^2) and forming 100-foot high cliffs, far too steep and unstable to climb (Figure 1.2).

Figure 1.2 Mt. Trident, Alaska, August 1963, looking to the northwest and showing the black 100-foot-thick lava flows erupted in the 1950s. The Valley of 10,000 Smokes (V) is seen one third of the way down the left edge. Katmai Pass (KP) is just to the south.

As we hike up along the western edge, I am thinking about the awesome power of Nature and about where we should run if the ground starts shaking. What's the quickest way out of here? We think the odds of an eruption today are very low, but they are not zero.

I reflect on the lengths to which scientists are willing to go in order to collect quality data. Are we crazy? The volcano actually did explode just three years later.

A large cloud of volcanic gases is blowing southward from the summit, dimming sunlight. As we climb around to the north side of the new peak, we find the wind is blowing so hard that the gases are pushed close to the ground, leaving us clean air to breathe as long as we remain standing (Figure 1.3).

Figure 1.3 Robert Decker near the north side of the summit of Mt. Trident looking at the smoking fumaroles.

As we climb higher and higher up the new peak, the pungent odor of **sulfur dioxide** is getting stronger. Too much sulfur dioxide mixed with water in your nose or lungs makes **sulfuric acid**. Not a healthy combination.

As we climb to the peak, we see more and more fresh sulfur deposited around gas vents—**fumaroles**—which are getting very close together. Thousands of these are the source of the gas cloud we had seen from below. We try to measure the gas **temperature** where it comes out of the ground (Figure 1.4), but it drives our thermometer off scale.

Figure 1.4 Decker measuring the temperature of a fumarole on Mt. Trident. In the background, center-left, you can see the infamous Valley of 10,000 Smokes, formed during the massive eruption of Mt. Katmai/Novarupta in 1912, the largest volcanic eruption in the 20th century.

The ground is too hot to sit on. Even my feet, safely inside my thick-soled Peter Limmer climbing boots, are feeling hot. The power of volcanoes up close truly is impressive. I find myself wondering whether scientists can ever understand volcanoes well enough to predict their eruptions, to save lives, and to minimize economic damage.

Living More Safely With Earthquakes

That was August 1963. In March 1964, the world's second most powerful earthquake ever recorded, with a magnitude of 9.2, ruptured the ground for more than 500 miles (800 km) along a fault stretching to the northeast from well south of Mt. Trident, past Kodiak Island, and all the way to Valdez, Alaska, which is 40 miles east of Anchorage (Figure 1.6).

The ground shook violently for four minutes. I know from personal experience that ground shaking for 20 seconds during an earthquake seems like an eternity. Four minutes! A lot can happen in that amount of time. Water waves—a tsunami—as high as 330 feet inundated the coast of Alaska. The ground moved permanently as much as 38 feet in some

Figure 1.5 The author examining a bread crust bomb erupted from Mt. Trident. As this 7-foot high block of lava was ejected into the air, the outside chilled, forming a crust while the inside continued to expand.

locations. Landslides and destroyed buildings were widespread, but thankfully population density was low. Only 139 people were killed, most due to the water waves. In 1976, a hundred times smaller earthquake of magnitude 7.5 in densely populated Tangshan, China, is reported to have killed 655,000 people in 16 seconds of severe ground shaking.

When I arrived back in Anchorage in June, I could see where five blocks of stores along the north side of Fourth Avenue had sunk 12 feet (Figure 1.7).

While I was eating dinner on the top floor of the Anchorage Westward Hotel, the waiter, with eyes as big as saucers, explained to me how everything in that large room slid from wall to wall as the building swayed back and forth for four minutes. The awesome power of the Great Alaskan Earthquake of 1964 made Mt. Trident seem pretty insignificant. By studying earthquakes, I wondered, could we predict them? Could scientists help us build communities that would be less vulnerable to Nature's wrath?

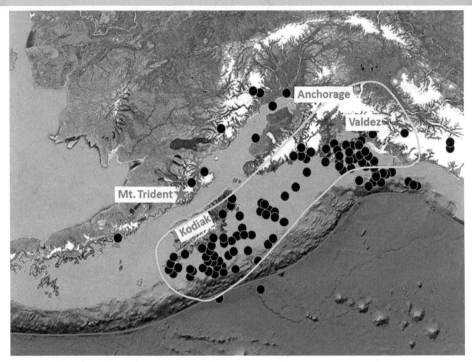

Figure 1.6 The Great Alaskan Earthquake of 1964 broke the ground at depth from Valdez to southwest of Kodiak Island. Red line is the Aleutian trench. Yellow line shows the area of major uplift. Circles show the locations of major aftershocks, dominantly at depths of 12 to 22 mi (20 to 35 km).

In June, 1965, I returned to Anchorage on my honeymoon. My wife and I had come to study earthquakes in the vicinity of Mt. Trident and their relationship to volcanism, but now I was a graduate student. I had decided to dedicate my life to studying earthquakes and volcanoes and to learning how we humans could live more safely with these extremes of Nature.

Katmai National Monument had been established in 1918 to preserve the area for the study of volcanism as a result of the Griggs expedition and several articles published in National Geographic. I was fortunate to spend three summers there in the 1960s and five more in the 1980s, studying earthquakes associated with volcanism in what became, in 1980, Katmai National Park, known to most people for the many pictures taken of bears gobbling salmon swimming up Brooks Falls to spawn in Brooks Lake (Figure 1.8).

The Great Alaskan Earthquake of 1964 was a wake-up call for America. After ten years of study, volumes of reports, a devastating magnitude 6.7 earthquake in San Fernando Valley north of Los Angeles in 1971, strong interest from Senator Alan Cranston (California), and

Figure 1.7 Five blocks along the north side of Fourth Avenue in Anchorage, Alaska, settled more than 12 feet during the Great Alaskan Earthquake of 1964.

years of scientific leadership by MIT professor Frank Press, who became Jimmy Carter's Science Advisor, earthquake hazard reduction was becoming a national priority.

On January 1, 1975, I was appointed Chief of the Branch of Seismology, a group of 140 scientists and staff at the United States Geological Survey in Menlo Park, California, right next to Stanford University. In the 1970s, scientists from around the world were discovering physical changes that were occurring days to months before major earthquakes. On February 4, 1975, local political leaders in Haicheng, Liaoning, China, ordered evacuation of dangerous buildings hours before a magnitude 7.3 earthquake severely damaged this city of one million people, saving countless lives. One tipoff to this earthquake was that snakes began to come out of hibernation during the winter, possibly because of warming of the ground due to rising water levels due to increasing pre-earthquake strain of the ground. Earthquake prediction is a high priority in China, where a majority of people live in poorly-constructed buildings that are likely to be destroyed during major local earthquakes.

There were many other tantalizing observations of possible earthquake precursors from around the world, but none was measured sufficiently well to allow us to understand the physics involved. In early December, 1975, my colleague Bob Castle handed me a preliminary plot of ground deformation[12] northeast of Los Angeles to present at our earthquake advisory panel meeting the next day. Could this deformation be a precursor to the next big southern California earthquake on a fault that had not slipped since 1857? The panel, chaired by Frank Press, agreed that it was entirely possible.

Consequently, in 1976, I was invited to brief the Presidential Advisory Group on Anticipated Advances in Science and Technology. Right after my presentation, Edward Teller, of hydrogen bomb fame, turned to Vice President Rockefeller and exclaimed that the possibility of predicting earthquakes was a very promising new advance in science and technology. The ponderous wheels of Washington began to turn a little faster, at least in regard to earthquakes. The Earthquake Hazards Reduction Act of 1977 (Public Law 95-124) established the National Earthquake Hazard Reduction Program. The Branch of Seismology, which I headed, became the Branch of Earthquake Mechanics and Prediction. Excitement was high. We were convinced that earthquakes just might be predictable if we could only trap several of them within dense networks of instruments so that we could map out the physical extent and accurate timing of these apparent precursors. We had already gotten pretty good at mapping out specific regions where earthquakes were most likely to

Figure 1.8 Catch of the Day, a famous photo by Thomas D. Mangelsen, at Brooks Falls in Katmai National Park, Alaska.

occur—regions known as seismic gaps, where no earthquakes had occurred for decades to centuries. Now, careful scientific observations and research might provide a way to save many lives.

Very soon, however, we began to realize that the physics of earthquake prediction might be easier than the sociology. In the midst of a research program in which you are trying to figure out reliable ways to predict earthquakes, how do you warn people that a major earthquake might occur tomorrow that could kill them, but you are only 5% certain? What action is appropriate if a devastating event is possible within a certain time interval but not highly likely? This made me reflect on our decision that it was okay to explore an active volcano up close, when we thought the likelihood of an eruption was low. Many volcanologists have done that. A few, however, did not return.

If we had credible information that an earthquake might occur, we realized that we had no choice but to communicate what we knew to the people at risk as clearly as we could, but how should we craft our message to make the information most useful? How could we help people prepare to make critical decisions rapidly if a prediction was issued?

We put instruments along many faults where earthquakes were highly likely, hoping to record at least one within a dense network of stations, but to our frustration, no earthquakes came out to play. They were apparently not on the same schedule as federal legislation.

We also found that good scientists, when faced with very sketchy data suggesting that many people could die, sometimes had trouble remaining completely objective. Each of them had a heart. They cared about the people at risk, and they seemed ready to take action even when the evidence was less than compelling. So we formed earthquake prediction review committees to help individual scientists to be as objective as possible before making critical decisions that might either save lives or cause unwarranted disruption.

Within a decade, however, with no large earthquakes occurring near our new instruments, we began to move more of the funding from prediction to hazard reduction— to making communities more resilient for when earthquakes do strike. The possibility of predicting earthquakes drove the establishment of the National Earthquake Hazard Reduction Program, but in America where most people live in wood-frame homes and well-engineered larger structures that generally survive earthquakes well, we decided that we could get more bang for the buck by reducing hazards and learning how to live more safely with major earthquakes.

Scientists still operate many thousands of instruments for locating earthquakes, monitoring potential precursors, and measuring deformation of the ground as it stores strain, which will be released in a future earthquake. Modern instrumentation is much more precise and covers much larger regions than possible in the 1970s, but predicting the moment when that strain will be released is still elusive.

A devastating earthquake did occur in 1989, the magnitude 6.9 Loma Prieta Earthquake, that severely damaged parts of San Francisco and Oakland, even though the epicenter was located 44 miles (70 km) to the south. A year later, a committee of scientists, using new methodology, determined that there was a 67% likelihood that a truly devastating earthquake would occur within 30 years along the Hayward Fault that courses through Fremont, Hayward, San Leandro, Oakland, Berkeley, Richmond, and San Pablo in the eastern part of the San Francisco Bay region in California. When it happens, this will likely be the most expensive earthquake in US history because of the population density. To educate the public about this grim likelihood, I was able to create, produce, and distribute, throughout the San Francisco Bay Area, 3.3 million copies of a 24-page magazine, with versions in English, Chinese, Spanish, and Braille, explaining the likelihood of this major earthquake and how to prepare for it.[13] There are many practical actions to take that will reduce the hazards. It was very gratifying for me to be able to advise people based on sound scientific research.

Twenty-five years later, the magazine has been forgotten by most people. This earthquake has still not yet happened—but at some point, it will happen, and while many preparations have been made to reduce losses, it will nonetheless be the most expensive earthquake so far in US history.

Living More Safely With Volcanic Eruptions

Meanwhile, volcanologists at the Hawaiian Volcano Observatory, established in 1912, have become rather successful at predicting volcanic eruptions in Hawaii,[14] where I first worked in 1967. Elsewhere, I had the opportunity to install the first seismograph on Mt. St. Helens in 1973 as part of a prototype volcano surveillance network that we were installing from Alaska, through the Cascade volcanoes in Washington and California, to volcanoes throughout Central America. A magnitude 4.2 earthquake occurred under Mt. St. Helens on March 20, 1980, attracting volcano experts from far and wide. Steam began venting a week later. The first eruption, on May 18th, was the result of a massive landslide from the fractured north slope, triggered by a magnitude 5.1 earthquake. All eruptions after that were successfully predicted.

One of my former field assistants, David Harlow, led the successful prediction of the eruption of Mt. Pinatubo in the Philippines in 1991, the largest volcanic eruption since the 1912 eruption of Mt. Katmai, just northeast of Mt. Trident in Alaska (Figure 1.1). Predicting the eruption of Pinatubo was not easy. It took substantial instrumentation to locate and monitor the earthquakes under the volcano, to measure ground deformation and changes in gas emissions, and to monitor and evaluate observable changes. It took close coordination with local disaster experts and with an American general who was in charge of nearby US Clark Air Force Base. The base was evacuated just before the eruption, was damaged by the eruption, and was abandoned just after the eruption. This successful prediction took a handful of scientists working around the clock, following their intuition honed by years of experience with other volcanic eruptions that they were not able to predict. Their agony and their ecstasy is captured in an excellent documentary "In the Path of a Killer Volcano: The Eruption of Mount Pinatubo," available on YouTube.[15]

Active volcanoes are now monitored around the world by local volcanologists with help from volcano experts from many other countries. Felt earthquakes and minor eruptions often occur weeks to months before an awakening volcano becomes dangerous to inhabitants of the region. Scientists are getting quite good at reacting to these early signs of activity. We are learning how to live more safely with volcanic eruptions.

Hazard maps have now been prepared for most potentially active volcanoes throughout the United States. These maps illustrate, for example, the paths of likely mud flows from Mt. Rainier into the densely populated Seattle, Washington area and airborne hazards along the Aleutian Island volcanic arc from Alaska to Russia—one of the busiest airline routes in the world.

Science, when done carefully, can save lives, can reduce losses, and can allay fears. I tell the foregoing story to emphasize that it is only since the Great Alaskan Earthquake of 1964 that scientists have made rapid progress in learning how to live more safely with earthquakes and volcanoes. I emphasize my role because I suspect that my decades of experience with earthquakes, volcanic eruptions, public policy, and public safety prepared me to notice a significant enigma in climate science and to undertake the careful and thorough investigations that led to the science described in this book. What a joy it is to work on something that makes the world safer for everyone and to be able to work outside, around the world, trying to understand Nature!

Living More Safely With Climate

The geologic record is replete with examples of geologically sudden changes in volcanism and associated changes in climate over decades, centuries, and millennia. For example, as noted in the Overview, studies of traces of ancient volcanic activity in dated Greenland ice cores have revealed that from 100,000 to 10,000 years ago, the world suddenly popped out of a major ice age 25 times within one, two, or three decades and then each time drifted back into the ice age over centuries to millennia. These cycles occurred, on average, once every 3600 years. Such major and geologically sudden change in climate had a substantial effect on the evolution of *Homo sapiens*, forcing our ancestors to find ways to adapt, often by migrating. Thus, our scientific studies about how to live more safely with volcanic eruptions are useful for understanding both long-term changes in climate and, as it turns out, short-term changes in weather observed in recent years. These effects on weather will be discussed in Chapter 9.

I retired in 1998 and moved to Jackson, Wyoming, to enjoy Nature by hiking, climbing mountains, rafting rivers, and skiing, and to play my guitar and piano. In 2006, however, Nature called me back into volcanology. While working on a book on an unrelated topic, I

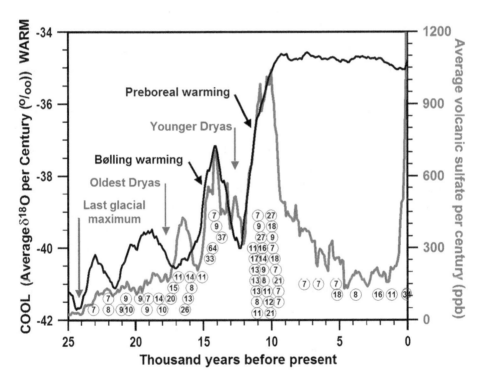

Figure 1.9 Average temperatures per century (black) increased at the same time as the amount of volcanic sulfate per century (red). The greatest warming occurred when volcanism was more continuous from year to year, as shown by the blue circles surrounding the number of contiguous layers (7 or more) containing volcanic sulfate. It was this continuity over two millennia that finally warmed the world out of the last ice age. Data are from the GISP2 drill hole under Summit, Greenland. Periods of major warming are labeled in black. Periods of major cooling are labeled in blue.

stumbled on very clear evidence that the end of the last ice age, from 11,750 to 9,375 years ago, coincided exactly with massive volcanic eruptions in Iceland (Figure 1.9).

At first, it didn't make sense to me that these eruptions could have ended an ice age because all volcanologists and climatologists understand that major explosive volcanoes typically cool Earth by nearly one degree Fahrenheit for up to three years. Such cooling has been observed clearly after most large, explosive eruptions in written history. Griggs described such cooling following the eruption of Mt. Katmai in 1912.[16] How could volcanoes cause both major cooling and major warming? It didn't add up, but the data were good, and good, reliable data are the closest things to truth in science.

For many months, I studied the layer-by-layer record of volcanic **sulfate** concentrations and the **oxygen isotope proxy** for temperature. That record had been collected by a fine team of scientists. The data had been analyzed carefully, and the possible errors seemed very small. When I showed these data to a leading climatologist, he said "Well, clearly there is something wrong with those data because volcanoes cause cooling," but the data were reliable. I can now say that the problem was with our understanding of climate.

After a life of climbing active volcanoes and trying to reduce the impact of natural hazards, I had a very clear gut feeling that getting to the bottom of this substantial enigma might lead to very important new insights into climate. The scientists who published these data did not seem concerned about them, but for me they were a red flag. I became obsessed. I decided to put aside almost everything else in my life in order to concentrate on trying to understand what was going on.

A major advantage of retirement is that I could concentrate on my own passions. There is no commuting, no committees to attend, no proposals to write, no promotion panels to please, no required training classes, very few interruptions, and no one to tell me regularly that the ideas I was evaluating were stupid. With the Internet, I had full access to scientific knowledge. I was free to follow my own intuition. The only caveat has been to step back regularly to be sure that I remain well grounded in trustworthy scientific observations.

I read thousands of scientific papers. I thought carefully about the foundational principles of climate change. I questioned in detail many assumptions made by many scientists, and increasingly, a few of these assumptions began to appear less sound. They did not stand up to thoughtful scrutiny, yet most scientists had considered them to be unarguable fact for nearly 150 years.

I studied many aspects of atmospheric chemistry, atmospheric physics, **thermodynamics**, **quantum mechanics**, and numerous other related fields. Being retired, I was free to follow my curiosity as to how volcanoes could cause both major cooling and major warming. The quest for understanding was fun, but for more than eight years, I went to bed exhausted and confused, waking up early the next morning full of new ideas, refreshed and ready to move forward. I was thrilled to make many new discoveries and even thrilled to discover a little later that some of my new discoveries were simply wrong. Fervent belief in your ideas drives you to work harder, but belief can be an obstacle to good science, and humility helps to keep you honest.

In science, it can be very important to tolerate ambiguity, to let things simmer, to accept possibilities as working hypotheses without having to jump to conclusions that could very well be incorrect. In life, we all need to reach conclusions so that we can decide how to move forward. Often there are time pressures. There is always the need to get on with it, to get it done so we can move on to the next problem to be solved. Scientists often draw conclusions based on their best understanding of common assumptions, based on widely accepted ideas. But when there are reliable data that raise hints of possible problems with widely held ideas, it takes time to become comfortable with new possibilities. The "whack-a-mole" scientist smashes new ideas the minute they pop up their ugly heads. The careful scientist walks a fine line between not jumping to conclusions and being productive. Intuition based on experience plays the major role in navigating along this fine line. Intuition is your brain whispering to you. Do you make time to listen?

Discovering a More Likely Cause of Global Warming

The evidence for volcanism in the ice layers under Summit, Greenland, consists of sulfate deposits. Sulfate comes from sulfur dioxide, megatons of which are emitted during each volcanic eruption. At first, I thought that the warming was caused by the sulfur dioxide, which is observed to absorb solar energy passing through the atmosphere.[17] My thinking was influenced by greenhouse warming theory, which assumes that carbon dioxide causes global warming because it is observed to absorb **infrared** energy radiated by Earth as it passes upward through the atmosphere and is then thought to re-radiate it back down to the surface, thus causing warming. The sulfur dioxide story, however, just wasn't adding up quantitatively.

Eventually, after publishing two papers that developed this story, I came to realize that sulfur dioxide was actually just the "footprint" of volcanism—a measure of how active volcanoes were at any given time. The real breakthrough came when I came across a paper reporting that the lowest concentrations of stratospheric ozone ever recorded were for the two years after the 1991 eruption of Mt. Pinatubo, the largest volcanic eruption since the 1912 eruption of Mt. Katmai. As I dug deeper, analyzing ozone records from Arosa, Switzerland[18]—the longest running observations of ozone in the world, begun in 1927 (Figure 8.15 on page 119)—I found that ozone spiked in the years of most volcanic eruptions but dropped dramatically and precipitously in the year following each eruption. There seemed to be a close relationship between volcanism and ozone. What could that relationship be?

The answer was not long in coming. I knew that all volcanoes release hydrogen chloride when they erupt, and I also knew that chlorine from man-made **chlorofluorocarbon** compounds had been identified in the 1970s as a potent agent of stratospheric ozone depletion. From these two facts, and a third one, I deduced that it must be the depletion of ozone by chlorine in volcanic hydrogen chloride—and not the absorption of solar radiation by sulfur dioxide—that was driving the warming events that followed volcanic eruptions. The third fact in the equation was the well-known interaction of stratospheric ozone with solar radiation.

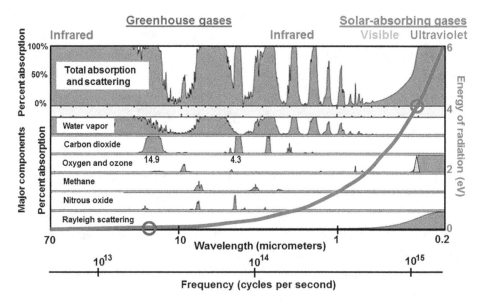

Figure 1.10 When ozone is depleted, a narrow sliver of solar ultraviolet-B radiation with wavelengths close to 0.31 μm (yellow triangle) reaches Earth. The red circle shows that the energy of this ultraviolet radiation is around 4 electron volts (eV) on the red scale on the right, 48 times the energy absorbed most strongly by carbon dioxide (blue circle, 0.083 eV at 14.9 micrometers (μm) wavelength. Shaded grey areas show the bandwidths of absorption by different greenhouse gases. Current computer models calculate radiative forcing by adding up the areas under the broadened spectral lines that make up these bandwidths. Net radiative energy, however, is proportional to frequency only (red line), not to amplitude, bandwidth, or amount.

The **ozone layer**, at altitudes of 12 to 19 miles (20 to 30 km) up in the lower stratosphere, absorbs very energetic solar **ultraviolet radiation**, thereby protecting life on Earth from this very "hot," DNA-destroying radiation. When the concentration of ozone is reduced, more ultraviolet radiation is observed to reach Earth's surface, increasing the risk of sunburn and skin cancer. There is no disagreement among climate scientists about this, but I went one step further by deducing that this increased influx of "super-hot" ultraviolet radiation also actually warms Earth.

All current climate models assume that radiation travels through space as waves and that energy in radiation is proportional to the square of the amplitude of these waves and to the bandwidth of the radiation, i.e. to the range of wavelengths or frequencies involved. Figure 1.10 shows the percent absorption for different greenhouse-gases as a function of wavelength or frequency. It is generally assumed that the energy absorbed by greenhouse-gases is proportional to the areas shaded in gray. From this perspective, absorption by carbon dioxide of wavelengths around 14.9 and 4.3 micrometers in the infrared looks much more important than absorption by ozone of ultraviolet-B radiation around 0.31 micrometers. Climate models thus calculate that ultraviolet radiation is relatively unimportant for global warming because it occupies a rather narrow bandwidth in the solar spectrum compared to Earth's much lower frequency, infrared radiation. The models neglect the fact, shown by the red line in Figure 1.10 and explained in Chapter 4, that due to its higher frequency, ultraviolet radiation (red circle) is 48 times more energy-rich, 48 times "hotter," than infrared absorbed by carbon dioxide (blue circle), which means that there is a great deal more energy packed into that narrow sliver of ultraviolet (yellow triangle) than there is in the broad band of infrared. This actually makes very good intuitive sense. From personal experience, we all know that we get very hot and are easily sunburned when standing in ultraviolet sunlight during the day, but that we have trouble keeping warm at night when standing in infrared energy rising from Earth.

These flawed assumptions in the climate models are based on equations that were written in 1865 by James Clerk Maxwell and have been used very successfully to design every piece of electronics that we depend on today, including our electric grid. Maxwell assumed that **electromagnetic** energy travels as waves through matter, air, and space. His wave equations seem to work well in matter, but not in space. Even though Albert Michelson and Edward Morley demonstrated experimentally in 1887 that there is no medium in space, no so-called **luminiferous aether**, through which waves could travel, most physicists and climatologists today still assume that **electromagnetic radiation** does in fact travel through space at least partially in the form of waves.

They also erroneously assume that energy in these imagined waves is proportional to the square of their amplitude, which is true in matter, but cannot be true in space. They calculate that there is more energy in the broad band of low-frequency infrared radiation emitted by Earth and absorbed by greenhouse gases than there is in the narrow sliver of additional high-frequency ultraviolet solar radiation that reaches Earth when ozone is depleted (Figure 1.10). Nothing could be further from the truth.

The story got even more convoluted by the rise of quantum mechanics at the dawn of the 20th century when Max Planck and Albert Einstein introduced the idea that energy in light is quantized. These quanta of light ultimately became known as **photons**. In order to explain the photoelectric effect, Einstein proposed that radiation travels as particles, a concept that scientists and natural philosophers had debated for 2500 years before him. I will explain in Chapter 4 why photons traveling from Sun cannot physically exist, even though they provide a very useful mathematical shorthand.

Max Planck postulated, in 1900, that the energy in radiation is equal to vibrational frequency times a constant, as is true of an atomic oscillator, in which a bond holding two atoms together is oscillating in some way. He needed this postulate in order to derive an equation by trial and error that could account for and calculate the observed properties of radiation. Planck's postulate led to Albert Einstein's light quanta and to modern physics, dominated by quantum mechanics and quantum electrodynamics. Curiously, however, Planck didn't fully appreciate the far-reaching implications of his simple postulate, which states that the energy in radiation is equal to frequency times a constant. He simply saw it as a useful mathematical trick.

As I dug deeper, it took me several years to become comfortable with those implications. It was not the way we were trained to think. It was not the way most physicists think, even today. Being retired turned out to be very useful because I could give my brain time to mull this over. Gradually, it began to make sense. The take-away message for me was that the energy in the kind of ultraviolet radiation that reaches Earth when ozone is depleted is 48 times "hotter" than infrared energy absorbed by greenhouse gases. In sufficient quantities, it should be correspondingly 48 times more effective in raising Earth's surface temperature than the weak infrared radiation from Earth's surface that is absorbed by carbon dioxide in the atmosphere and supposedly re-radiated back to the ground.

There simply is not enough energy involved with greenhouse gases to have a significant effect on global warming. Reducing emissions of greenhouse gases will therefore not be effective in reducing global warming. This conclusion is critical right now because most of the world's nations are planning to meet in Paris, France, in late November 2015, to agree on legally binding limits to greenhouse-gas emissions. Such limits would be very expensive as well as socioeconomically disruptive. We depend on large amounts of affordable energy to support our lifestyles, and developing countries also depend on large amounts of affordable energy to improve their lifestyles. Increasing the cost of energy by even a few percent would have major negative financial and societal repercussions.

This book is your chance to join my odyssey. You do not need to have majored in science or even to be familiar with physics, chemistry, mathematics, or climatology. You just need to be curious and be willing to work. You also need to be willing to think critically about observations, and you may need to reevaluate some of your own ideas about climate. You will learn that there was a slight misunderstanding in science made back in the 1860s that has had profound implications for understanding climate change and physics today. It took me many years of hard work to gain this insight, and I will discuss that in Chapter 4. First, however, we need to look at some fundamental observations that cause us to wonder: Could the greenhouse warming theory of climate change actually be mistaken?

CHAPTER 2

COULD CLIMATE CHANGE
SCIENCE BE MISTAKEN?

"New scientific ideas never spring from a communal body, however organized, but rather from the head of an individually inspired researcher who struggles with his problems in lonely thought and unites all his thought on one single point which is his whole world for the moment."

—Max Planck, 1936

Everyone has an opinion about "climate change"—it will render Earth unlivable; it's a non-issue; it's caused by humans; it's entirely natural; the scientists must be right; the scientists are simply pandering for funding. Many liberals insist we must act now—many conservatives claim that reducing greenhouse-gas emissions will bankrupt our economy. For lack of logical arguments, many resort to attacking the motives of the opposition. The climate debates have turned rancorous. The world has become polarized over whether or not to reduce greenhouse-gas emissions in order to mitigate climate change that may or may not be occurring, and, if it is occurring, may or may not be human-caused.

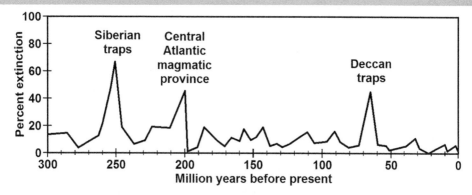

Figure 2.1 The three major peaks in extinction of species were contemporaneous with three of the largest eruptions of flood basalts forming the Siberian traps in Siberia, the Central Atlantic Magmatic Province, and the Deccan traps in India.

Climate is critically important. There is nothing on Earth that affects us all so comprehensively. All living things depend on it, and they must either adapt to climate change or die. At the end of the Permian period, 252 million years ago, climate changed so radically that 96% of all marine species and 70% of all terrestrial vertebrate species vanished (Figure 2.1)[19] in what is known as the Great Permian Extinction.[20] At the same time, one million cubic miles (4 million km³) of **basaltic** lava—the kind you see flowing

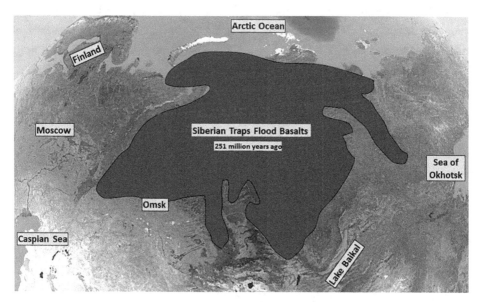

Figure 2.2 The Siberian Traps Province covers an area in Siberia that is equivalent to 87% of the contiguous United States or all of Western Europe.

down hillsides in Hawaii and Iceland—covered 2.7 million square miles (7 million km²) of Siberia (Figure 2.2).[21] That is an area equal in size to 87% of the contiguous United States or all of Western Europe! If volcanism like this were to recur today, there would be no one left to have an opinion about climate change. Fortunately, such events are rare, occurring on average once every 23 million years.[22] The most recent flood basalt event was much smaller, forming the Columbia River Basalts, which covered 63,000 square miles (163,000 km²) of Washington and Oregon around 16 million years ago.

The climate change we face today is much more manageable, but still substantial. Species are currently becoming extinct at 1000 times the normal background rate, although this likely has more to do with habitat destruction caused by our rapidly increasing human population than with climate change per se.[23] Average global **temperature** has increased approximately 1°F (0.6°C) since 1965.[24] Minimum temperatures rose 12°F (6.7°C) on the Antarctic Peninsula,[25] sea level has risen 9 inches since 1870,[26] and the weather often seems more severe, bringing more devastating floods to some regions and drought to others.

For those affected by the storm surge during Hurricane Sandy, known locally as "Superstorm Sandy," climate change suddenly became a significant concern. Hurricane Sandy's effects were local, and people had to adapt in crisis mode, but a drought in California affects the entire United States. The current California drought is the worst, based on tree-ring studies, in more than 1200 years,—i.e., since almost 700 years before Columbus "discovered" the New World.[27] Over the last century, the United States has grown to depend on California for much of its food: 100% of all US almonds, 99% of artichokes, 99% of walnuts, 97% of kiwis, 97% of plums, 95% of celery, 95% of garlic, 89% of cauliflower, 71% of spinach, 69% of carrots, etc.[28] As the drought continues, availability of these foods will decrease, prices will increase, and California's economy will suffer greatly.

Adapting to Change

All living things spend every moment of their lives adapting to change in their physical and social environments. We humans continually adapt to changing weather, deciding what to wear or what to do based on a number of factors including what we hope to accomplish that day, our own observations and experience, the observations and actions of others, or the predictions of meteorologists. The need to adapt is immediate, and the penalty for not adapting adequately is also immediate. Most of the time, the penalty may not be all that significant, but if a tornado is headed our way, we could die if we do not take appropriate shelter immediately. At that point, we do not have time to debate what actions are most appropriate. This is when training and practice really pay off.

We each have limited time and resources for adapting to change. Whether and how we choose to adapt depends on our understanding of the problem, how immediate the threat is, our assessment of the risk, how important it is to us to put ourselves at risk, if we perceive that there are effective actions we can take with or without others, whether we think we can get by without taking action, and so on. All animals, including human beings, are capable of taking executive action, of deciding what actions to take, when, where, how, and why, based on past experience, current knowledge, their assessment of their goals, their priorities, their world view. We do this all day, every day, balancing many conflicting needs.

The reason we all have opinions about climate change is that we have a deep sense that climate change could potentially be a threat to our quality of life, if not to life itself. Thus, each of us must ultimately decide whether current and predicted changes could be a problem for us and whether we should take action to adapt to them. Climate is, by definition, long-term trends in weather, so climate does not typically pose an immediate threat. We have time to think, question, and discuss. What, precisely, is the problem? What is the solution? Can actions that I take have any positive effect? How much will it cost for me to take action? How does this action compare to other pressing actions I need to take? Should I pool my own efforts with those of others? How urgent is the need? We approach these questions based on our world views, which are based in part on our genetic propensities toward liberal or conservative approaches to life. Since the threat is not imminent, believing that climate change is not a serious problem is the easiest way out. The Pew Research Center found that in the United States during 2013-2014, 77% of scientists were convinced climate change is a very serious problem while only 33% of non-scientists agreed.[29] What do the scientists understand that the non-scientists don't? Are the scientists right?

What Is Science?

Science is the most objective method we have for fact finding and fact checking based on observation and on both logical induction and logical deduction that can be independently verified, and it is thus self-correcting over time. Most scientists seek to understand how Nature works. Observation is the ultimate authority in science, but observations vary in quality and reproducibility. Scientists seek the most accurate observations, and then they try to understand what caused the observed phenomenon—the well-known inverse problem: given a result, determine the cause. Typically, however, there are many possible causes. Which one is right? Which possible cause best explains all the observations? Just because two things happen simultaneously does not necessarily prove that one caused the other.

Scientists typically propose hypotheses, from which, if the hypotheses are substantiated by observation and experiment, they then develop theories—logical ways of explaining observations. The more observations a theory can explain, the more scientists tend to accept the theory. Over time, a well-established theory is used to define what observations should be made, but here there is a danger of circular reasoning. Investigations based on a particular theory can be designed in a manner that tends to prove that theory. But what if that theory is not correct?

A fundamental goal in science, and especially in physics, is to write a mathematical equation that describes the observations. This equation can then be used to infer what would happen under different circumstances. In this way, equations become central to designing and engineering products that dominate our modern life. Sometimes an equation is derived empirically, by trial and error, but an equation is valued most if it can be derived from other equations and theories that appear to be well established.

In converting an observation to an equation, however, the physicist must make assumptions based on what seems logical within his/her world view. Our individual views tend to correspond closely with those of our professional peers, especially in science, but each component of any world view typically started with the interpretations and assumptions of a single scientist.

Once an equation is formulated, it becomes useful to mathematicians, who do not necessarily care about the pertinent physics. Mathematics is a symbolic shorthand for logic, yet there are many things that make perfectly good logical sense in mathematics but not in physics. Imaginary numbers, for example, defined by the square root of minus one, make perfect sense in mathematics and are extremely useful, but they are not physically real. They are, as their name indicates, imaginary. In mathematics, preceding a number with a minus sign is quite logical and a negative number is a real number. But does it make sense physically to put a minus sign in front of a particle or in front of matter to define antiparticles or antimatter?

Max Planck and Albert Einstein—two of the greatest physicists of the 20th century—developed by 1905 the concept of a **quantum** of energy. That concept led to modern physics, which is dominated by **quantum mechanics**. Both had excellent physical intuitions, but when the mathematicians took over, quantum mechanics became less physical. Any introductory textbook on quantum mechanics will tell you that its subject does not make intuitive, physical sense. Never mind that. Just get over it, as we all have had to do. As Richard Feynman is purported to have put it, "Shut up and calculate!"[30] Einstein had a famous ongoing debate with Nils Bohr from 1926 to 1949[31] until he reluctantly agreed that quantum mechanics was internally consistent, as is any good mathematics. What disturbed Einstein most about quantum theory was its total renunciation of all minimal standards for realism. Both Planck and Einstein, the fathers of quantum mechanics, went to their graves upset over the fact that quantum mechanics does not seem to make physical sense. Mathematics is required to do good physics, but remember that the ultimate goal is to explain real things, and real things are only a rather narrow subset of possible mathematical solutions.

Science seeks ultimate knowledge. Science also seeks perfection, but it is a human endeavor, and as such is subject to human limitations. What makes science work, and what makes society work, too, is that most humans want to be appreciated and respected not only by those they love and respect, but also by their peers. Most of us try hard to do what we feel is right both logically and ethically within the constraints of the circumstances. A few cheat, but that doesn't detract from the reality that most people work hard and with integrity in order to make a difference. A scientist's work is always subject to review by other scientists. Maintaining credibility and professionalism is especially important to scientists.

What is even more important is that science evolves. It usually builds on the past in small increments, and each increment is typically done by an individual. Over time, all scientists build houses of cards with related observations, experiments, theories, and mathematics. Typically, these structures are refined and strengthened, becoming quite robust. Some even succeed in solving major problems, but they are still basically houses of cards. There is always the possibility that some observations or assumptions made long ago could turn out to be incorrect. When this happens, the card houses can tumble, leading to a major revolution in scientific thinking.

Is There a Role For Belief in Science?

Science is built on verifiable observations made and analyzed by humans who are motivated to work hard on causes they believe in. Belief is perhaps the greatest motivator, the greatest spur to action for all humans—belief in our own ability to make a difference; belief in an idea or a line of logic; belief that taking certain actions will help us best adapt to change, earn us more money, a better quality of life, or greater success in life. Beliefs define how we see the world. Beliefs form our personal guidebook and also our moral compass. We use our beliefs daily to set priorities, to decide what actions to take, what information to ignore, and what is morally right.

Our beliefs are built on experience, observation, knowledge learned from others, and our cultural heritage but also on our emotions, intuition, gut feelings, and biases. Many people find religious belief empowering. Religious beliefs are traditional, based on human experience and on interpretations from long ago. They cannot be fact checked, but they can be powerful motivators for individuals, parishes, and societies.

Belief also has its drawbacks, however. One often reads in news reports the phrase "Scientists believe that...." Strictly speaking, belief is actually antithetical to the scientific method because once something is believed, it becomes fixed, canonized, written in stone, and is difficult to change. Ideas that cannot be changed belong in traditional religions, not in science. The great strength of science is in its flexibility, its ability to change on the basis of new information, new hard data. Without that flexibility, science would be unable to perform its fundamental task of seeking out unbiased truth. Once a scientific hypothesis or theory is believed, it cannot change, and it therefore becomes an article of faith rather than being a guidepost on the pathway to truth. In science, everything must remain on the table, always subject to revision on the basis of new and better evidence.

Unfortunately, most non-scientists, and far too many scientists as well, fail to understand this critical constraint on scientific theories. Perhaps a better alternative to the phrase "Scientists believe that..." would be "Scientists suspect that..." or "Scientists think that according to available evidence, this seems most likely." While scientists ideally seek verifiable observations and seek a perfect understanding of the natural and social worlds, their interpretations inevitably involve some assumptions, some steps in logic that may not be as thoroughly fact-checked as other conclusions but have nonetheless proved very useful over time and have become part of scientists' store of working hypotheses. For this reason, it is extremely important to look for and to question assumptions, and particularly, beliefs. Most scientists, however, are busy building on these assumptions so that they can be more productive, getting on with our scientific understanding, and they tend to disparage others who question ideas that form the foundation of their own work. Here, again, belief has trumped suspicion. Within scientific communities there is social pressure to conform.

The Intergovernmental Panel on Climate Change (IPCC), organized under the United Nations, is an interesting case in point. The panel was formed in 1988 to bring the growing scientific consensus about climate change to the attention of political leaders in order to get consensus on appropriate political action to mitigate a perceived growing problem. While consensus is the stuff of politics, however, debate is the stuff of science. Thousands of scientists gave up major amounts of time from their research to document consensus, welding a body of knowledge together to show that climate change is occurring and that it must be caused by observed increases in emissions of **greenhouse gases**. People who seriously questioned the growing consensus were not invited to participate, while scientists whose conclusions best illustrated the growing consensus were catapulted to leadership roles. This is the process of group-think, and it is very human. The impressive, detailed reports by the IPCC put out every four years or so were compiled by hard-working scientists who apparently believe sincerely that they are saving the world from impending climate disaster. For their efforts in service to mankind and to Nature, the IPCC was awarded a Nobel Peace Prize. If the scientific consensus turns out to be correct, the IPCC's efforts will go down in history as a glowing example of how scientists and politicians can work together to save life on Earth. If it doesn't, it might well serve as a glowing example of how belief can be misapplied in scientific endeavors.

Scientists and Politicians Have Worked Well Together

The IPCC grew out of a very successful collaboration of scientists and politicians that did, in fact, stop human-caused global warming, as described in the next chapter. In 1974, Mario Molina and F. Sherwood Rowland published a paper[32] showing that **chlorofluorocarbon** gases (CFCs), used widely as spray-can propellants, refrigerants, solvents, and foam blowing agents because they were so chemically inert, could be broken down by ultraviolet solar **radiation** during particularly cold spells in the lower **stratosphere**. Molina and Rowland showed that through **catalytic** reactions, a single atom of chlorine could destroy more than 100,000 molecules of ozone, as explained in Chapter 6. The **ozone layer**, primarily at altitudes from 12 to 19 miles (20 to 30 km) above Earth's surface, absorbs most ultraviolet-B radiation from Sun. This is the radiation that causes sunburn, skin cancer, etc., but it is also used by our bodies to create vitamin D. When the ozone layer is depleted, more of this dangerously "hot" radiation reaches Earth's surface, warming Earth.

Scientific concern over ozone depletion led, in 1985, to adoption of the Vienna Convention for the Protection of the Ozone Layer. Discovery of the large ozone hole over Antarctica in that same year[33] led to the **Montreal Protocol** on Substances that Deplete the Ozone Layer, becoming effective at the beginning of 1989, which limited the manufacture of CFCs. The Vienna Convention and the Montreal Protocol were the two most successful treaties of all time, ultimately ratified by all members of the United Nations. In this case, the science linking CFCs to ozone depletion was clear, the danger of ozone depletion to life on Earth was clear, the scientists involved worked hard to educate political leaders about the problem, and the formation of the Antarctic ozone hole demonstrated that the science was probably right, adding a strong sense of urgency. Furthermore, most CFCs were manufactured by one major chemical company that had alternative chemicals available. The financial burden of reducing emissions of CFCs was manageable both by producers and by consumers, and principles were established for developed countries, which caused the problem, to help undeveloped countries adapt. The Vienna Convention and the Montreal

Protocol are major successes made possible by scientists and politicians working together to identify and solve a major environmental problem.

Where Am I Coming From?

For as long as I can remember, I have been obsessed with trying to figure out how things work. I enjoy a wide variety of puzzles, but I am passionate about trying to understand practical things that have real effects on my life and on the world. As a child, I took broken things apart in the hope of being able to fix them, and when I put them back together, there were often parts left over. Sometimes my curiosity paid off. Many times it didn't, but I was always learning. From an early age, I decided that the only bad experiences in my life were ones from which I could not learn. This approach to life was certainly reinforced when, at the age of 12, I watched, up close, as my best friend for five years died in a horrific accident. This approach helped when I only got to visit my mother on Sundays for a couple of years, standing outside the window of her room at a tuberculosis sanitarium. One can only try to make the best of any situation.

Partly as a way of dealing with these experiences, partly from watching my father, and partly by temperament, I developed an ethic early on of working very hard over long hours. I am happiest when I am able to focus most of my energy, around the clock, on something to which I am sincerely dedicated. If there is something I feel is important that needs to be done, I like to roll up my sleeves and get on with it. You only live once, so why not make the most of life?

As a field geoscientist, I am most committed to quality observations about things I can see, touch, and experience both macroscopically and microscopically, with and without instruments. To me, physics has always been about the physical, how Nature works. While I have taken numerous graduate level courses in mathematics, I do not delight in advanced theoretical physics because it all depends on the assumptions of those who have come before. While theoretical physics is of great value, and while I sincerely appreciate the work of many great minds in theoretical physics, I often wonder about the line between physical physics and brilliant mathematics. I have a deep abiding belief that while the physical details may get very complex, new, more fundamental understanding based on direct observation and careful experiment will bring us closer to a relatively simple theory of everything. This book is a hopeful step in that direction.

Where Am I (and You) Going?

In 2006, having been retired for 8 years, and while working on something totally unrelated to climate science, I came across some excellent data on the Internet collected from ice cores taken from the **GISP2** drill hole beneath Summit Greenland (Figure 1.9 on page 13).[34] Analysis of these cores gave a **proxy** for surface temperature at the time when the snow turned to ice and a measure of the amount of volcanic **sulfate** within the layers, each of which represents approximately 2.5 years of time. These very detailed data showed that the amount of volcanic sulfate, i.e. the amount of volcanism, was greatest and most continuous from 11,750 to 9,375 years before the present, precisely the time when the world finally warmed out of the last ice age. That suggested to me that volcanism could have caused the warming, yet it is well-known to all volcanologists and climate scientists that major explosive **volcanoes** cause global cooling for several years. How could volcanoes cause both cooling and warming?

I climbed my first erupting volcano, Mt. Trident in the Katmai region of the Alaskan Peninsula, at the age of 19 and specialized for more than a decade in studying earthquakes related to volcanic eruptions in Alaska, Iceland, Hawaii, Washington, Oregon, California, Guatemala, El Salvador, and Nicaragua. The evidence that volcanoes might cause both cooling and warming was an enigma that piqued my curiosity. How could that be? I sensed intuitively at the time that understanding this enigma could lead to a breakthrough in our understanding of climate, although I had no idea how that understanding might change.

What started out as curiosity soon turned into an adventure, an odyssey, a quest that still dominates my life. Building on my extensive background in geology, physics, volcanology, seismology, and plate tectonics, I started reading in depth about climatology, paleoclimatology, atmospheric chemistry, atmospheric physics, radiation physics, quantum mechanics, quantum electrodynamics, etc. Being retired and well connected to the Internet, I was free to follow my curiosity wherever it led. I started out trying to understand this enigma, but ended up critically evaluating all the observations and assumptions made in support of the greenhouse warming theory of climate change. I have waded through more than 10,000 scientific papers, dozens of websites, numerous books, many textbooks, and many video courses. When ideas began to emerge that questioned conclusions of the IPCC, I was free to let them simmer. I did have to do reality checks regularly to be sure I was not slipping off the edge of the rational world. After all, if you end up disagreeing with most other scientists, statistically there is a good chance that you are catastrophically wrong. Nevertheless I just kept delving into the details, looking for the most reliable observations.

CHAPTER 3

WHY DID GLOBAL TEMPERATURES STOP INCREASING IN 1998?

"The essence of human beings is not in their ability to reason but in their ability to rationalize."

—Anonymous

A ir temperatures are changing all the time. We often consult a thermometer or a weather report to decide what to wear or whether certain activities today would be appropriate or even safe. We observe and have learned to adapt to the reality that temperatures vary from lows around sunrise to highs in the afternoon, that they vary daily, weekly, monthly, and by season, that they vary with cloud cover, and that they decrease with increasing altitude and with increasing latitude. How is it possible to combine all these data to calculate an average annual global surface **temperature** that will be useful for understanding and quantifying climate change over years to decades?

The first sealed thermometer that was suitable for measuring air temperatures was made in Italy in 1641.[35] Since 1659, temperatures in central England have been measured routinely.[36] Most nations have meteorological organizations that measure and record air temperatures many times per day by means of instruments typically placed at an elevation between 4.1 and 6.6 feet (1.2 to 2 m) above the ground in order to reduce the effect of direct solar heating of the ground. In 1978, Phil Jones and others at the Climatic Research Unit (CRU) of the University of East Anglia in England began compiling these

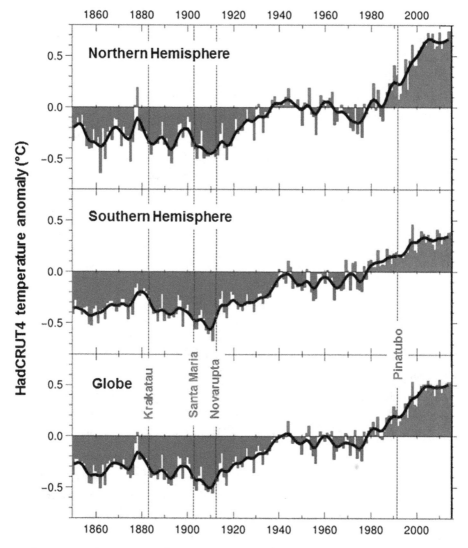

Figure 3.1 Average annual temperature anomaly increased twice as much since 1970 in the northern as in the southern hemisphere. Cooling normally follows major volcanic eruptions near the tropics, shown in red, except for Novarupta (Mt. Katmai) which was much farther north.

data.[37] They chose 1850 for a start date, in order to be able to include a meaningfully representative number of stations. Today, their network includes 5583 land surface stations around the world. Meanwhile, the Hadley Centre of the United Kingdom Met Office had been compiling sea surface temperature records. Jones and his staff[38] combined these two data sets in 1986 into what became the Hadley-CRU Temperature dataset, or HadCRUT, currently in version 4.[39,40] They calculated monthly temperature averages at each station and then averaged all the monthly station averages in each 5 degrees of latitude by 5 degrees of longitude box over Earth's surface, ultimately calculating a global average.

To measure climate change, we only need to know the average change in temperatures over time. Since land stations are in different local circumstances, at different elevations and latitudes, measured at different times of day, and measured and averaged in slightly different ways, Jones decided to calculate average monthly temperatures at each station and then to compare these to average monthly temperatures at the same station over the 30-year period from 1961 through 1990, during which good station coverage was maintained.[41] The resulting values, expressed as monthly temperature anomalies, or deviations from the monthly averages, are plotted in Figure 3.1 for the northern and southern hemispheres and for the globe as a whole.[42] Note that the warming anomalies in the northern hemisphere since 1970 have nearly twice the values of the warming anomalies in the southern hemisphere.

Temperature Trends

Long before the invention of thermometers, it was well known from historical records that the world was as warm as, or possibly warmer than, it is now during the Medieval Warm Period from approximately 950 to approximately 1250 AD.[43] This was the time when the Vikings colonized southern Greenland and grapes were being grown in northern England. Then, around 1350, the globe began to cool by as much as one degree centigrade into what is known as the Little Ice Age, with the lowest temperatures thought to have occurred around 1650, 1770, and 1850.[44] The HadCRUT4 compilation of thermometer data since 1850 in Figure 3.1 shows 1862 as the coldest temperature anomaly in the northern hemisphere (0.65 °C) and 1911 as the coldest temperature anomaly in the southern hemisphere (-0.67 °C). These cold temperatures were influenced by the major **explosive** volcanic eruptions named in red in the figure. I will discuss these and other types of eruptions and their influence on temperature in Chapter 8.

Perhaps the most interesting observation in Figure 3.1 is that recent global warming occurred in two pulses, increasing approximately 0.5°C from 1911 to 1940, and then increasing approximately 0.6°C from 1970 to 1998. Except for these two warming pulses, global temperature anomalies remained relatively constant around 0.05°C from 1940 to 1970 and around 0.56°C from 1998 to 2014. How reliable are these observations? The HadCRUT4 global data set is shown again in Figure 3.2, in which it is compared to the three other principal analyses of global temperature anomalies. How were these other three data sets compiled, and how do they compare with HadCRUT4?

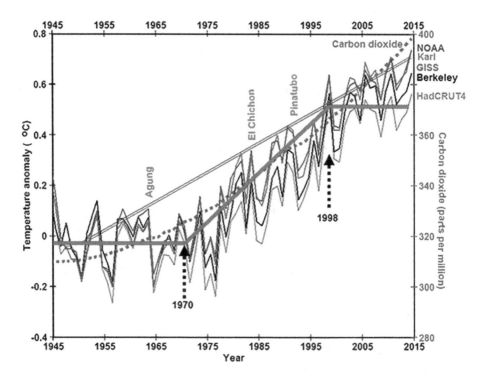

Figure 3.2 The red line shows schematically the average change in surface temperature anomalies for the four land-and-sea-based major datasets (HadCRUT4, Berkeley Earth, GISS, and NOAA). Temperatures changed very little form 1945 to 1970, rose sharply until 1998, and changed very until 2014. The increase in temperature anomalies in 2014 and 2015 is quite likely caused by ozone depletion resulting from the eruption of Bárðarbunga volcano in central Iceland from August 2014 to February 2015 that extruded 33 square miles of basalt, the largest basaltic eruption in Iceland since 1783. The dotted blue line is the atmospheric concentration of carbon dioxide that has just kept increasing. The fuchsia double line shows the temperature trend calculated by Karl et al.,[17] which does not seem to fit the data.

In the 1980s, James Hansen and others[45] at NASA's Goddard Institute for Space Studies (GISS) developed a surface temperature analysis with special concern for urban warming and other local human influences on the global temperature data.[46] They chose a different 30-year base period for calculating temperature anomalies, in this case from 1951 through 1980.

In the early 1990s, the NOAA National Centers for Environmental Information developed the Global Historical Climatology Network (GHCN), integrating land-based data from more than 20 sources and applying "a common suite of quality assurance reviews"[47] to all data. Sea surface temperatures were added from an extended, reconstructed sea surface temperature analysis based on the International Comprehensive Ocean-Atmosphere Data Set.[48] They chose a base period from 1971 through 2000 for calculating temperature anomalies.

In 2010, Richard Muller, a physicist at the University of California, Berkeley, founded a non-profit group called Berkeley Earth[49] to take an independent and rigorous look at the climate data reported by the three organizations just described. They collected 14.4 million average monthly temperature observations from 14 databases, including 44,455 instrument sites on land. They also addressed concerns that had been voiced by climate skeptics about the warming effect of major heat-generating urban development near some of the thermometer sites, about any possible risk of bias in data selection, and about poor station quality. They concluded that land temperatures have increased by 1.5°C since 1750 and by 0.9 ±0.05°C from the 1950s to the 2000s.[50] They integrated these land data with a modified version of the Hadley Centre Sea Surface Temperatures (HadSST) to arrive at the data plotted as the black line in Figure 3.2.

Note that the four different datasets and four different analytical techniques show broad agreement from year to year and in long-term trends, even though each uses a different base period to calculate the temperature anomalies. Their common, long-term trends are summarized by the red line with two prominent inflection points at 1970 and 1998. Annual average global surface temperatures were relatively constant from 1945 to 1970, rose steadily and dramatically by about 1°F (0.54°C) from 1970 to 1998 (0.34°F or 0.19°C per decade), and have remained at a warmer but relatively constant temperature since 1998. Cooling is observed for a few years after each of the three volcanic eruptions labeled in red, as will be discussed in more detail in Chapter 8.

Reviewing these data, most people would likely agree that the best thermometer measurements available show that the world did warm by a little more than 0.9°F (0.5°C) from 1970 to 1998, but that outside that time interval, temperature anomalies have tended to remain relatively constant, although at two different levels, since 1945. The four datasets similarly indicate that there was an earlier warming of approximately 0.9°F (0.5°C) between 1911 and 1940 (shown for one dataset in Figure 3.1).

A recent re-analysis of the global surface temperatures by Karl et al.[51] that carefully considers data biases concludes that there has been no significant hiatus in global warming since 1998 and that temperatures increased 0.113°C per decade from 1950 to 1999 and 0.116°C per decade from 2000 to 2014. These rates are shown by the double fuchsia line in Figure 3.2. Note that all four data sets fit the solid redline much more closely, showing little change from 1945 to 1970, rapid change from 1970 to 1998, and little change since 1998. In an effort to show little change in an upward trend since 1998, Karl et al.[17] averaged temperatures going all the way back to 1950 (double red line), ignoring the rapid increase in temperature shown clearly between 1970 and 1998 by all major datasets (red line).

The Stark Divergence Between Increasing Carbon Dioxide Concentration and Invariant Temperature

The greenhouse warming theory is built on the assumption that an increase in the atmospheric concentration of **carbon dioxide** will cause a corresponding increase in average global temperature.[52] One current estimate suggests that a doubling of the carbon dioxide concentration is likely to increase global temperatures somewhere between 4 and 9°F (2.2 and 4.8°C).[53] Another says between 2.7 and 8.1°F (1.5 and 4.5°C).[54] The dashed blue line in Figure 3.2 shows the annual average atmospheric concentration of carbon dioxide, in **parts per million** (ppm). The more recent portion of the curve, from 1957 to the present, was measured at Mauna Loa Observatory[55] in Hawaii. The earlier portion was measured in ice cores at Law Dome[56] in Antarctica, and its values were adjusted to match up with the beginning of the Mauna Loa data in 1957.

Clearly, the blue carbon dioxide concentration curve shows an inconsistent relationship to the red trend line of the average global surface temperature anomaly. The red line has two inflection points around 1970 and at 1998, respectively after and before periods in which no significant change in temperature was observed. There are no such sudden changes in the slope of the blue carbon dioxide curve, nor is there any known or proven process that would cause such sudden changes in the carbon dioxide trend. Since 1998, there has been a stark divergence between rapidly increasing atmospheric carbon dioxide concentration and non-increasing temperature anomalies. Moreover, unlike the temperature anomalies, the rate of increase in carbon dioxide concentration has been accelerating through time. From 1945 to 1970, it only rose 0.6 ppm per year; from 1970 to 1998, it rose 1.5 ppm per year; and since 1998, it has been rising at 2 ppm per year. Given these many differences, therefore, the most straightforward and logical interpretation of Figure 3.2 is that carbon dioxide is not the primary cause of global warming.

Rationalizing the Role of Natural Variations in Climate

Most climatologists today are convinced beyond all doubt, however, that increases in concentrations of carbon dioxide and other **greenhouse gases** are indeed the primary cause of global warming. More than 50 research papers, all listed on my website,[57] were published by the end of 2014, each one trying to explain the stark divergence in trends between rapidly increasing concentrations of carbon dioxide and relatively constant average temperatures, a divergence that is in its 18th year as of this writing. Climatologists refer to the constant temperatures since 1998 as a "global warming hiatus," with the implication that it is only a temporary phenomenon that will, in time, be reversed as global warming "comes roaring back with a vengeance," as one researcher put it. Another confidently states that "in the end, global warming will prevail."[58]

In order to explain this "hiatus," climate scientists invoke both "external" natural variations that add energy to, or remove energy from, the climate system, and "internal" natural variations that simply redistribute energy over time in the oceans and in the atmosphere.

The external changes include a prolonged minimum in the 11-year solar sunspot cycle, variations in Sun's energy output, changes in land use, increased **aerosols** caused by humans or by volcanism, decreased water vapor in the lower **stratosphere**, and changes in the **radiative forcing** of greenhouse gases.

Internal factors include changes in ocean currents that move heat around the globe, movement of heat to greater depths in the ocean, strengthening of trade winds, and changes in salinity in the subpolar North Atlantic. Changes in ocean currents discussed include the El Niño Southern Oscillation (ENSO), the Pacific Decadal Oscillation (PDO), the Interdecadal Pacific Oscillation (IPO), Atlantic Meridional Overturning Circulation (AMOC), the Atlantic Multidecadal Oscillation (AMO), strengthening of Indo-Pacific ocean currents, and Antarctic Bottom Water (AABW) formation. Those interested in details regarding these winds, currents, and oscillations can find them all clearly described in Wikipedia.

The conclusions reached by the various papers discussing the global warming "hiatus" are varied, suggesting a lack of consensus. Earlier papers tend to be dismissive of the plateaued trend in temperature and confident of a swift return to warming, but more recent ones show less confidence. Three possible causes stand out, but there is little evidence of a convergence of opinion around any of them. One is the role of both volcanic and man-made aerosols in reflecting away incident solar **radiation**. Another is the transfer of heat from the shallow to the deep ocean, although evidence for such a transfer is unclear, and underlying transfer mechanisms are poorly understood. A third is the strengthening of the so-called Walker circulation in the equatorial Pacific, resulting in La Niña conditions, in which the northeasterly trade winds increase, moving surface water westward and cooling the eastern Pacific through upwelling. Other papers cite a weakening of solar output, and one even invokes astronomical effects from Earth's motion through the galaxy. The overall impression is one of grasping at straws in the hope of finding a satisfactory explanation for the enigmatic "hiatus," and of a lack of a clear convergence upon one or two strong candidates for such an explanation.

Three other prominent features are also noteworthy. First, although at least one paper acknowledges that the "hiatus" could potentially erode confidence in the climate models, none goes so far as to question the capability of these models to make long-term forecasts. Second, none of these papers ventures the heretical thought that there might be something incorrect in greenhouse warming theory and that some hitherto unidentified cause might

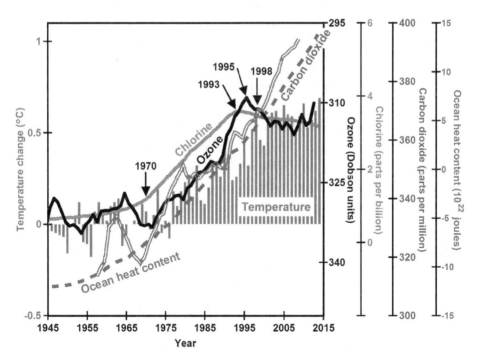

Figure 3.3 The increase in tropospheric chlorine (green line), caused by manufacturing of chlorofluorocarbon gases, led to increased ozone depletion (black line), which led to increased temperature (red bars). All these trends reversed from 1993 to 1998.

have been responsible for the dramatic global warming episode of the late 20th century. Third, none of these papers addresses why the warming began gradually between 1965 and 1970 and why, in contrast, temperatures stopped increasing abruptly in 1998.

The Ozone Depletion Theory of Global Warming

These three trends—the gradual onset of warming after 1965, rapid warming from 1970 to 1998, and the abrupt termination of rapid warming in 1998—can be explained simply and directly by ozone depletion, which was clearly observed by scientists to be occurring at the same time. In the 1960s, **chlorofluorocarbons** (CFCs) became popular for use as refrigerants, spray-can propellants, solvents, and foam blowing agents because they do not interact with most other chemicals. By 1970, a wide variety of products in spray cans had become available with CFCs as propellants. Emissions of these human-manufactured, chlorine-bearing gases began increasing by 1965 (green line, Figure 3.3).[59]

By 1970, total column ozone, measured poleward of the tropics, became depleted by as much as 50%, especially in the southern hemisphere, resulting in the well-known **Antarctic ozone hole** that reaches its peak development during mid to late local winter (black line).[60]

The **ozone layer**, extending from 12 to 19 mi. (20 to 30 km) above Earth's surface, absorbs high-energy ultraviolet-B radiation from Sun, thereby raising its temperature. This radiation is "hot" enough to burn human skin and to cause skin cancers and cataracts. When the ozone layer is depleted, more ultraviolet-B radiation than usual is able to reach Earth's surface, cooling the ozone layer and warming Earth.

As I noted previously, Mario Molina and F. Sherwood Rowland[61] discovered, in 1974, that CFCs can be broken down by **ultraviolet radiation** in very cold **polar stratospheric clouds** located below the ozone layer during late winter and early spring, thereby releasing chlorine atoms. Each atom of chlorine can destroy on the order of 100,000 molecules of ozone through **catalytic** reactions that I will discuss in Chapter 5. After discovery of the Antarctic ozone hole[62] in 1985, scientists and politicians worked efficiently together under the Vienna Convention for the Protection of the Ozone Layer to develop the **Montreal Protocol** on Substances that Deplete the Ozone Layer, which became effective on January 1, 1989. By 1993, subsequent to implementation of this protocol, increases in CFC emissions stopped; increases in ozone depletion stopped by 1995,[63] and increases in global temperatures stopped by 1998 (Figures 3.1 to 3.3). Increasing emissions of CFCs appear to have caused the gradual increase in temperature from 1965 to 1970. Decreasing emissions of CFCs beginning in 1993 appear to have stopped increases in ozone depletion by 1995 and further increases in temperature by 1998. Because CFC concentrations continue to decrease slowly, further increases in temperature due to CFCs are not anticipated.

Annual average ozone concentrations have remained depleted since 1998 by approximately 4% in northern mid-latitudes compared to pre-1970 concentrations. The resulting increased influx of ultraviolet-B radiation continues to increase ocean heat content (Figure 3.3, fuchsia double line)[64] because it penetrates tens of meters into the ocean,[65] from which depth the energy is not radiated back into the atmosphere at night.

Ozone depletion ranges from zero in the tropics to more than 50% during late winter-early spring in polar regions. Similarly, global warming also ranges from zero in the tropics to many degrees in polar regions, suggesting a cause-and-effect relationship. The greatest global warming ever recorded, 12°F (6.7°C), was measured from 1951 to 2003 on the Antarctic Peninsula,[66] and this was the greatest warming observed for this region in 1300 years.[67] There was also significant warming in West Antarctica[68] and in the Arctic.[69] Amplification of warming temperatures in the polar regions,[70] which has been widely observed, is fully consistent with ozone depletion theory because the greater the ozone depletion, the greater the warming. It is very difficult to explain Arctic amplification with greenhouse warming theory.[71]

Figure 3.1 shows that there was twice as much warming from 1965 to 1998 in the northern hemisphere (1.4°F or 0.8°C) as in the southern hemisphere (0.7°F or 0.4°C). This could be a reflection of the fact that in the northern hemisphere, which contains 68% of the global land area and 88% of the global population, the elevated influx of ultraviolet-B radiation is initially being absorbed in the lower troposphere by low-level ozone generated by pollution, which is many times greater in the northern hemisphere than in the southern. This, in turn, could perhaps explain the observation that although Earth's average surface temperature increased while ozone depletion was increasing, it is remaining relatively constant while ozone depletion gradually declines.

From the foregoing, it appears likely that global warming since 1965 was caused by the observed depletion of the ozone layer. Since 1998, however, by which time the United Nations Montreal Protocol had led to substantially reduced emissions of CFC gases that were causing the ozone depletion, the world has not continued to warm, but remains warmer than it was before 1965 because ozone remains depleted.

Why has this not been acknowledged by most climatologists? Due to the way in which they calculate the energy contained in radiation (see the discussion in Chapter 1), they conclude that there is much more energy involved in the infrared radiation absorbed by greenhouse gases than there is in the ultraviolet-B energy that reaches Earth when ozone is depleted (Figure 1.10 on page 16). Personal experience suggests, however, that this is untrue. As noted, ultraviolet radiation feels hot, and it may even cause sunburn. Earth's infrared radiation, as experienced at night in the absence of sunlight, on the other hand, feels cold. The confusion over the nature of radiant energy has deep roots going back to 1865, and the debate has involved all the best minds in physics. In the next chapter, I will try to shed some light (as it were) on this problem.

Table 4.1 Higher frequency radiation contains more energy than, and is much hotter than, lower frequency radiation. Values shown, except for room temperature, are for the top of the radiation band so that, for example, frequencies for extreme ultraviolet radiation range from 30,000 to 2,998 THz, etc. The red highlighting shows frequencies, energies, and temperatures involved when ozone is depleted; the green highlighting shows frequencies, energies, and temperatures involved when greenhouse gases absorb terrestrial infrared radiation.

Radiation Band	Minimum Wavelength Nanometers	Maximum Frequency Terahertz	Maximum Energy Electronvolts	Color Temperature Kelvin	Effects
Gamma rays	0.001	3x108	1.24x106	2.9x109	Lethal even in small amounts
Extreme ultraviolet	10	30,000	124	290,000	Ionizes N_2, O_2, etc. Forms and heats *ionosphere*
Ultraviolet-C	100	2998	12.4	29,000	Dissociates O_2, SO_2, etc. Heats *stratosphere*
Ultraviolet-B	280	1071	4.43	10,300	Dissociates O_3, sunburn, vitamin D, skin cancer
Ultraviolet-A	315	952	3.94	9200	Skin cancer, fading of materials
Visible light	380	789	3.26	7630	Photosynthesis, dissociates NO_2, NO_3, HONO
Near infrared	750	400	1.65	3860	Begin absorption by water vapor
Short wavelength infrared	1400	214	0.886	2070	Absorption by water vapor
Mid-wavelength infrared	3000	99.9	0.413	966	Main absorption by greenhouse gases
Long-wavelength infrared	8000	37.5	0.155	362	Main absorption by greenhouse gases
	9804	30.6	0.127	296	Room temperature, 23°C, 73°F
Far-infrared	15,000	20	0.0827	193	
Microwave	100,000	3	0.0124	29	
Longwave AM radio	$1.08x10^9$	$2.79x10^{-7}$	$1.15x10^{-6}$	$2.7x10^{-3}$	

CHAPTER 4

DO WE REALLY UNDERSTAND THERMAL ENERGY?

"If you want to find the secrets of the universe, think in terms of energy, frequency, and vibration."

—Attributed to **Nikola Tesla**

E arth is both warmed and cooled by radiant thermal energy. It is warmed primarily by Sun's **ultraviolet radiation** and visible light that is absorbed as it flows through our atmosphere and as it reaches Earth's surface. It is cooled primarily by much cooler infrared radiation flowing from Earth's surface and atmosphere back into space. As long as the total thermal energy received from Sun by the Earth/atmosphere system is equal to the total thermal energy radiated back into space by that system, then long-term, average **temperature** just above Earth's surface should not change. If excess solar energy reaches Earth, however, or less infrared energy reaches space, then long-term, average temperature just above Earth's surface is likely to increase.

Most climate scientists think that when concentrations of **greenhouse gases** increase in the atmosphere, less energy from Earth escapes to space, and more is returned to Earth's surface, resulting in global warming. In this book, I propose that global warming results instead from the depletion of ozone, which allows more high-energy ultraviolet solar radiation than usual to reach Earth's surface. Most solar ultraviolet-B radiation (Table 4.1) is normally absorbed by the **ozone layer** 12 to 19 miles (20 to 30 km) above Earth in the lower **stratosphere**. This is the radiant energy that burns your skin and can cause skin cancer. If the concentration of ozone in the ozone layer is reduced (depleted), less ultraviolet-B radiation is absorbed by that layer, whereupon it cools, and more ultraviolet-B radiation than normal is observed to reach Earth, warming its surface, and increasing our risk of sunburn and skin cancer.

Before we can determine the relative effectiveness of greenhouse gases versus ozone depletion in causing global warming, we must first establish a clear understanding of what we mean by light, an electromagnetic field, energy, chemical energy, thermal energy, radiation, and temperature, and how thermal energy flows through matter, through the atmosphere, and through space.

In this chapter, I get to the heart of the physics and the **thermodynamics** of global warming. If physics scares you, be assured that you do not need a background in physics to understand, at least on a conceptual level, what is written in this chapter. There is no mathematics to decipher. The arguments are based on direct, everyday observations that anyone can make. If you still have trouble comprehending the details, just focus on the major conclusions and understand that there is strong support for them.

If you are a physicist, on the other hand, you might find this chapter challenging because I demonstrate a number of things that differ, to some extent, from your academic training. Understanding these differences may give you new insight into some of the most challenging problems in physics. As the name suggests, physics is about physical things and what is physically happening. To do good physics, we need mathematics to quantify potential causes of reliable observations. Yet mathematics can provide numerous ways to describe observations, and we need to be careful only to pick those mathematical solutions that are clearly physically relevant. Our goal is to understand what light and thermal energy are, how they actually propagate, how they are radiated, and how they are absorbed. The best physics is based on clear and unambiguous observations of Nature.

What Is Light?

Democritus (~460 to ~370 BC) suggested that all things in the Universe are composed of indivisible sub-components that later came to be known as atoms. He thought, for example, that light consists of a stream of solar atoms or particles. Aristotle (384 to 322 BC) hypothesized that light is a disturbance in a hypothetical element known as "aether" and is thus propagated as a wave. The Arab polymath Ibn-al Haytham (or Alhazen, 965 to 1040 AD) wrote the first treatise on optics describing reflection and refraction of light. He thought of light as traveling from source to viewer along straight, linear rays composed of particles.

In his 1630 treatise on light, René Descartes described reflection and refraction by modeling wave-like disturbances in an aether, a medium that he called a plenum. In 1704, Isaac Newton published his book *Opticks* developing a corpuscular hypothesis for light as a stream of particles. He reasoned that only particles could travel along such straight lines, or rays, observed for light. Robert Hooke (1635 to 1703) and Christiaan Huygens (1629 to 1695) developed the mathematics of light travelling through an enigmatic medium that became known as the **luminiferous aether**. In 1818, Augustin Fresnel[72] noticed that light could be polarized and concluded that light must therefore travel as transverse waves that oscillate perpendicularly to the direction of travel. He understood, however, that transverse waves can only propagate in matter, where the bonds holding the matter together provide the restoring forces that allow the waves to propagate. He therefore stressed that there must indeed be some form of luminiferous aether in space that somehow allows light to propagate. In 1849, Michael Faraday[73] introduced the concept of an electromagnetic field in air and space consisting of electric and magnetic waves oscillating in mutually perpendicular planes, with both planes being perpendicular to the direction of travel (Figure 4.1).[74]

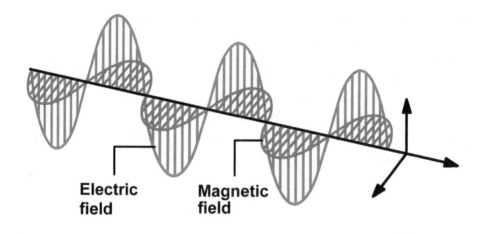

Electric field   **Magnetic field**

Figure 4.1 Faraday's visualization of an electromagnetic field propagating to the right, consisting of an electric field (red) oscillating perpendicularly to a magnetic field (blue), both oscillating perpendicularly to the direction of propagation.

By 1865, James Clerk Maxwell[75] had formulated a set of partial differential equations showing that electric and magnetic fields in space satisfy the wave equation when you think of **electromagnetic radiation** as transverse waves traveling at some velocity through some medium. For decades, physicists sought to discover what Fresnel's luminiferous aether was or to prove that it could not exist. Most physicists agreed that waves need to travel through some medium, but the search for the luminiferous aether was extensive and not successful.[76] A famous experiment by Michelson and Morley,[77] in 1887, finally convinced most physicists that such an aether does not exist and therefore that waves cannot propagate through space. Maxwell's equations, however, became not only highly respected, but they have been used very successfully to design every piece of electronics that has been invented to date. They must therefore have some validity. Physicists were left with no choice but to conclude that electromagnetic radiation is somehow different from waves in an aether.

In 1905, Albert Einstein[78] proposed that light could travel as particles of energy that he called "light quanta," and that ultimately came to be known as **photons**.[79] Einstein was trying to understand the photoelectric effect, in which electrons are released from the surfaces of certain polished metals when they are illuminated with deep blue to ultraviolet light. He likely reasoned that for a particle, an electron, to be given off, it might have been struck by another particle, a light **quantum** or a photon, although he did not express it precisely in this way. Since 1905, most physicists have adopted the concept of **wave-particle duality**,[80] whereby they think of light traveling either as a wave, à la Maxwell, or as a photon, à la Einstein, using whichever mathematical approach seems appropriate for the problem at hand.

We cannot see light. We only see the effects of light when it shines on matter as solids, liquids, or particles dispersed in air, such as water and dust particles that are illuminated when a searchlight is pointed up into the sky. What I find fascinating is that throughout the 2400 year history of distinguished natural philosophers and scientists thinking about light, we have tried to explain something we cannot see, light, in terms of something we can see, either waves or particles.

We cannot see the wind. We only see the effects of wind when it interacts with matter. As the poet Christina Rossetti (1830 to 1894) wrote "Who has seen the wind? Neither I nor you. But when the leaves hang trembling, the wind is passing through." Should we explain the wind in terms of waves or particles? We can think of wind as being reflected, dispersed, and even refracted by porous media. We could write wave equations to describe these phenomena, except that the fine scale of atmospheric turbulence would favor a statistical approach.

I also have a logical problem with wave-particle duality. If light is sometimes like A and sometimes like B, then is it not logical that light might actually be something different from both A and B?

I will explain, later in this chapter, observations suggesting that light cannot propagate either as waves or as particles, but for now, suffice it to say that I think most physicists today can agree that light is an electromagnetic field, which is what Maxwell sought to describe with his equations. Let us therefore consider next the nature of fields.

What Is an Electromagnetic Field?

In physics, a field is considered to be a physical quantity that has a specific value at each location in three-dimensional space that can change as a function of time. We typically cannot see a field unless we introduce some type of matter into it. A magnet, for example, generates a magnetic field with lines of force that can be visualized directly when we sprinkle metal particles on a piece of paper held close above, but not touching, the magnet (Figure 4.2).[81]

When a compass is brought near the magnet, the compass needle will line up along these lines of force. The lines have no physical reality; they simply illustrate, in a way that we

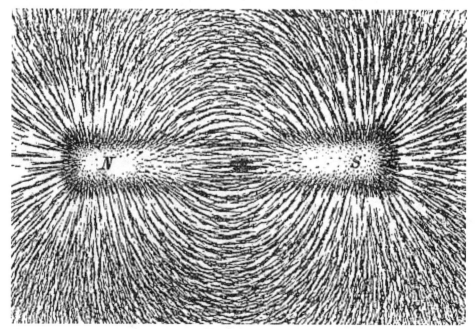

Figure 4.2 Iron filings on a sheet of paper held over a bar magnet with north (N) and south (S) poles line up to illustrate the existence and properties of a magnetic field.

can visualize, the directions of the invisible forces at each location in the three-dimensional field. The invisible lines connect the north pole of the magnet (N) with its south pole (S).

In the late 1700s to the early 1800s, a new type of field, an electric field, was discovered with the invention of the capacitor, an electronic component that is commonly used in most pieces of modern electronics. A simple capacitor consists of two metal plates arranged parallel to, but separate from, one another, one of which is connected to the positive terminal of a battery and the other to the negative terminal. The direct current from the battery's negative terminal drives a flow of electrons (i.e., a current) toward one of the plates that continues until the charge density on that plate equals that on the battery, at which point the current in the circuit slows to a stop. No electrons can flow between the two plates, but as negatively charged electrons accumulate on the one plate, they drive a flow of electrons away from the opposing plate, giving it a positive charge. If you were then to replace the battery with an alternating current source, this same process would occur, but with electrons being driven alternately toward one plate and then toward the other, thereby maintaining a continuous current in the circuit, although, as in the direct current example, no electrons can flow between the two plates.

It was discovered, however, that an electric field existed between the two plates as well as surrounding the wires connected to the alternating current source. Alternating current entails an oscillation from positive to negative at some rate, or frequency. Electricity supplied to a house typically oscillates from positive to negative and back to positive 50 or 60 times per second (50 or 60 cycles per second), generating an electric field around wires and other electrical components. Electric fields only form if the electric current that creates them is alternating, or oscillating at some frequency. The **frequency of oscillation** is a fundamental property of an electric field. Electric fields do not form about a wire carrying direct current that does not oscillate in any way. To repeat a very important point: an electric field does not, and cannot, exist unless there is some frequency of oscillation. Frequency is not only a fundamental property of an electric field, but it is a required property.

An electric field oscillating at some frequency induces a magnetic field oscillating at the same frequency as the electric field and perpendicularly to it, which in turn induces an oscillating electric field, and so on, forming a propagating electromagnetic field. Maxwell's equations for electromagnetism[82] calculate that the velocity of a disturbance propagating through this field is equal to the reciprocal of the square root of the product of two physical constants: the vacuum permittivity (the resistance to forming an electric field) and the magnetic permeability (the ability to form a magnetic field). In other words, the velocity of light is a function of how rapidly an electric field can induce a magnetic field, plus how rapidly a magnetic field can induce an electric field. This is a very short time, since the velocity of light equals 6.71 x 10⁸ miles per hour (299,792,458 meters per second), but the frequencies of oscillation can be very large numbers. In the visible spectrum, red light has frequencies of 380 to 484 THz (**terahertz**, or trillions of cycles per second), green light has frequencies of 526 to 606 THz, and violet light has frequencies of 668 to 789 THz. The full spectrum of electromagnetic frequencies ranges from well below 10^4 cycles per second to well above 10^{20} cycles per second, as shown in Figure 4.3.

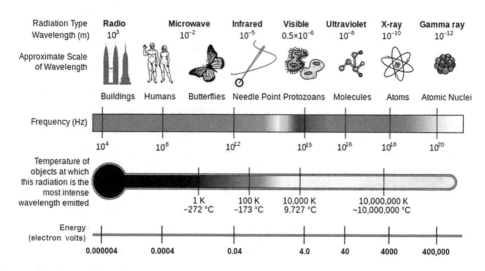

Figure 4.3 The electromagnetic spectrum extends over at least 14 orders of magnitude of frequency. Energy is equal to the frequency times the Planck constant. The temperature of objects for which the radiation at a given frequency is the most intense, the color temperature, similarly increases with frequency and energy.

Any material object within direct view of the source of any part of this electromagnetic radiation field is illuminated by the radiation, causing two physical properties to be displayed: frequency and amplitude. We perceive frequency as color, and we perceive amplitude as brightness. Each frequency is a distinct color that does not change with distance, and each color has a distinct amplitude—a brightness—that decreases in inverse proportion to the distance travelled. When radiation is absorbed by matter, those frequencies with which the bonds of the matter can resonate are absorbed, provided that the amplitudes of the incoming radiation are greater than those of the absorbing matter at corresponding frequencies. Those frequencies that determine the color on the surface appear to be reflected or scattered, but the nature of reflection and scattering is poorly understood, as is the nature of the energy that is apparently transmitted by the field. It may be that nothing travels through the field, but only that oscillation on one side of the field causes oscillation at a point in direct line of sight somewhere else in the field. There is still much about electromagnetic energy that we do not understand.

Thinking of light, and other forms of radiation, as an electromagnetic field might be confusing. Color and brightness are communicated across space via line of sight from source to receiver. At a given distance from the source, the amplitude, or brightness, is different for every color, or frequency of oscillation. With increasing distance, the color does not change, but the brightness decreases with (i.e., is inversely proportional to) the square of distance travelled. Remarkably, however, although there can be trillions of sources in a given region of space, neither their frequencies nor their amplitudes interact with each other except in the immediate presence of matter. Radio signals occur at the low frequency end of the electromagnetic spectrum (the left end of Figure 4.3). Millions of radio signals, each from a different station, can be transmitting simultaneously across space without interfering with one another except when the signals from two or more localities are emitting nearly the same frequency and the amplitudes of these signals at the receiver are somewhat similar.

Mathematically, we define the wavelength of light as the velocity of light divided by its frequency. We have become accustomed to calculating the wavelength of light, to describing light in terms of waves, and to plotting the spectrum of light as a function of wavelength, but this is mathematics, not physics. The concept of wavelength is simply not needed to describe light fully.

When I mention to physicists that electromagnetic radiation is simply frequency, they usually reply "Frequency of what? How can I visualize frequency?" Aye, there's the rub! We cannot see (visualize) light until it interacts with matter. The frequency of visible light appears to our eyes as color. Every shade of color in the rainbow has a unique frequency. As we look at the rich visual world around us, every molecule of colored matter is oscillating at some frequency. This frequency is transmitted through air and space. When we look in the direction of a molecule that is transmitting in the visual spectral range, the frequency that the molecule's oscillating bonds transmits causes parts of the rods and cones in our eyes to oscillate at the same frequency, sending neurologic signals to our brains that allow us to perceive that molecule as a particular color. We do not see light (frequency) as it travels toward us, but we do see frequency (color) when it causes that frequency of oscillation in our eyes.

A little further on in the discussion, I explain how heating matter increases the amplitudes and frequencies of oscillation of all the bonds that hold matter together. These bond oscillations induce an electromagnetic field that propagates away (radiates) from the surface of matter until it is absorbed by some other piece of matter, whereupon it increases the amplitudes of bond oscillations on the surface of that matter, thereby raising its temperature.

What Is Energy?

The answer to this simple question is far from simple. Richard Feynman, one of the best known physicists of the 20th century, wrote in his highly respected Lectures on Physics in 1963: "It is important to realize that in physics today, we have no knowledge of what energy is. We do not have a picture that *energy comes in little blobs of a definite amount*."[83] Feynman received the 1965 Nobel Prize in Physics for his contributions to the development of quantum electrodynamics, which is the study of how light (radiation) interacts with matter. Understanding quantum electrodynamics is closely related to understanding how solar radiation interacts with Earth's atmosphere and Earth's surface. Feynman's statement that energy does not come "*in little blobs of a definite amount*" contradicts what is assumed by all climate models and by all climatologists when they calculate **radiative forcing**. Radiative forcings are additive, they each lead to little blobs of energy of a definite amount.

Jennifer Coopersmith, in her book *Energy, the Subtle Concept*,[84] concludes that energy is "what makes things happen." Energy is "the 'go' of the Universe." Energy causes things to happen around us. No change is possible without utilizing some energy. Thus, far from being a thing "*in little blobs of a definite amount*," energy is actually a description of the condition or state of a thing—the 'go' of a thing. Physicists like to define energy as the ability to do work and to define work as force times displacement, but this definition does not apply to thermal or chemical energy. You get energy from digesting food, and this allows you to do work. Plants absorb energy from Sun, allowing them to grow. Dead plants contain energy that we can access by lighting them on fire. Fossil plants contain energy that we can utilize when we burn coal. The amount of energy that something has refers to its capacity to cause change. Higher (greater) energy typically makes more things happen in a shorter period of time.

A fundamental property of energy is that it is conserved, as was first stated by William Rankine in 1850 in the law of the conservation of energy.[85] Energy cannot be created or destroyed. It can only be converted from one form to another. This is known as the first law of thermodynamics—the study of thermal energy in motion. This law has been interpreted as meaning that the total amount of energy in the Universe is fixed, although it is strictly applicable only to thermodynamically isolated systems. Climate change is all about converting radiant energy in air and space to thermal energy stored in matter and vice versa.

What Is Thermal Energy?

When we heat something, it absorbs thermal energy, and it gets hotter, i.e., it reaches a higher temperature. What, then, is heat? How is heat stored in matter? How is heat related to temperature?

Temperature is a macroscopic physical property of gases, liquids, and solids—something we can sense and measure, and something to which all living creatures must adapt in order to survive. It is an intensive physical property, which means that it is not dependent on the size of the piece of matter. If a body of matter is divided, each piece will initially have the same temperature immediately following the division. This implies that macroscopic temperature is determined by something pervasive at the microscopic level, something that, at microscales, is going on everywhere all the time.

Scientists have observed that the bonds that hold particles of matter together are not rigid. The length of each bond oscillates in response to two opposing electrostatic forces—one that attracts the particles when they are relatively far apart, and another that repels them when they are too close together. The term "electrostatic" signifies invisible forces

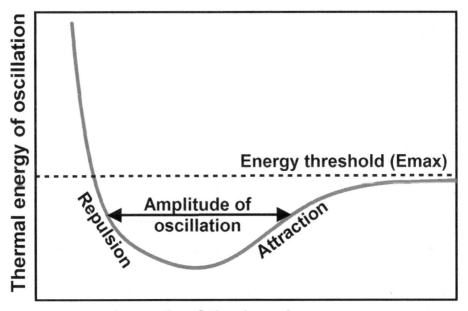

Figure 4.4 *The length of each bond holding matter together oscillates between electrostatic forces of repulsion on the left and electrostatic forces of attraction on the right. When energy equals or is greater than the threshold energy (Emax) the bond comes apart.*

caused by differences in the electric charges that are bound to the particles, (i.e., that do not generate an electric current or an electric discharge moving away from the particles). We experience static electricity when our hair stands on end, when we scuff our feet along certain carpets on days with very low humidity, or when plastic wrap clings to our hands.

The length of oscillation of the molecular bond shown in Figure 4.4 oscillates along the horizontal black line with arrowheads between a minimum value along the red line on the left, labeled "Repulsion," and a maximum value along the red line on the right, labeled "Attraction." As the thermal energy increases, the amplitude of oscillation increases, and the black line in Figure 4.4 rises and lengthens, until the amplitude becomes so large at the energy threshold (E_{max}) that the bond comes apart, causing the matter to melt or vaporize as its constituent atoms or molecules fly apart.

Such electrostatic oscillations are often compared to mechanical oscillations, such as a pendulum swinging back and forth. The longer the pendulum or the heavier the mass on the bottom of the pendulum, the lower the frequency of oscillation.

There are two very important differences between mechanical oscillators and an electrostatic atomic oscillator. As shown in Figure 4.4, electrostatic oscillators are asymmetric, with longer amplitude of oscillation on the attraction side than on the repulsion side. This means that as thermal energy increases, the mean amplitude (length of the bond) increases, which means that the material expands. It is widely observed that most solids, liquids, and gases expand when heated.

The second difference is that the electrostatic oscillator has essentially no friction, so that it can oscillate with constant amplitude indefinitely. The only way to add or subtract energy from such an oscillator is by **resonance** with a nearby oscillator. Resonance occurs when two oscillators near each other are oscillating at similar frequencies but with different amplitudes of oscillation. The oscillator with the lower amplitude will absorb amplitude from the oscillator with the higher amplitude until they both have the same amplitude at the same frequency. Since **amplitude of oscillation** is a function of energy of oscillation, as shown in Figure 4.4, thermal energy is shared and equalized, and will reach **thermal equilibrium** primarily via resonance. Thermal equilibrium means the condition, also known in physics as the state, in which thermal energy (heat) no longer flows spontaneously from one oscillator to another.

Resonance is all around us, and many examples are illustrated on YouTube, including sound breaking a wine glass,[86] tuning forks exchanging energy,[87] the Tacoma Narrows Bridge collapsing due to resonance in 1940,[88] and standing waves in a pipe.[89] Two dimensional examples include resonance in a membrane similar to the head of a drum,[90] and waves on the surface of water.[91]

Each of the bonds that hold matter together can resonate in several independent ways, known as **degrees of freedom**. Each degree of freedom has certain natural **frequencies of oscillation** that are dependent on the masses of the atoms and the strength of the electrostatic forces involved. It is the "thermal motions of the molecules, their bonds, vibrations, rotations, and excitations"[92] that store microscopic thermal energy within matter. Indeed, the capacity of a material to store heat is observed to increase with increasing number of degrees of freedom of oscillation within the material. This is, in theory, what makes greenhouse gases important. Greenhouse gases have three or more atoms, and thus have many more degrees of freedom of motion relative to each other than do molecules with only one or two atoms. Molecules of greenhouse gases can stretch symmetrically and asymmetrically, scissor, rock, wag, and twist in addition to the six degrees of freedom available to molecules with only two atoms.[93]

Thermal energy stored in the bonds that hold matter together is referred to in thermodynamics as **internal energy**. A body of matter possesses three types of energy: 1. macroscopic **potential energy** (the potential to be displaced in some direction, as by the force of gravity), 2. macroscopic **kinetic** energy (when actually being displaced in some direction), and 3. microscopic internal thermal energy (due to internal oscillations at the molecular level with no net movement of the body in some direction). The atomic-scale dimensions of these oscillators are very small, so their natural frequencies of oscillation at room temperature are very high, around 30.6×10^{12} cycles per second, i.e. 30.6 trillion cycles per second, or 30.6 terahertz (THz) (Table 4.1). When matter becomes warmer, the amplitudes of these oscillations increase at all frequencies, and the frequency with the largest amplitude also increases.

What Is Radiation?

Each atom in these atomic oscillators has some net charge that is continually in motion, oscillating at very high frequencies. As we have seen, the oscillations of charged particles on the surface of matter induce an electric field. The motion of charge also induces a magnetic field. The charge's motion within the magnetic field, in turn, induces an electric field, and so on, forming self-propagating electromagnetic radiation that spreads away from the surface of a body of matter.

As mentioned above, a field is something you cannot see, but it occupies three-dimensional space outside of matter. If you put a test charge somewhere in the field, you can detect and determine the properties of the field. Electromagnetic radiation consists of an interconnected electric and magnetic field that is self-propagating through air and space and that contains a broad spectrum of frequencies and amplitudes of oscillation generated by all the minuscule atomic oscillators located on the surface of the radiating matter.

When you listen to a radio station, you are utilizing a single frequency component of an electromagnetic field. The radio transmitter causes molecules on the surface of a radio antenna to oscillate at some specific frequency, such as 88.5 million cycles per second (megahertz) on your FM radio dial, which is National Public Radio in many areas. These oscillations on the antenna transmit an electromagnetic field, adding a high amplitude signal at this frequency to all the other electromagnetic signals out there. When you are within range of the radio station and you tune your radio dial to that frequency, the radio receiver is tuned to resonate at exactly 88.5 megahertz, receiving, by resonance, some of the amplitude at that frequency while ignoring all signals at all other frequencies. What is actually transmitted through air or space is frequency with some amplitude. There are thousands of frequencies being transmitted all the time by radio stations, television stations, cellular telephones, two-way radios, Wi-Fi equipment, Bluetooth devices, etc. They all co-exist within the electromagnetic field and each can be received independently of the others by an appropriate receiver tuned to resonate at the specific frequency of interest. Their signals do not interfere with each other except in locations where signals with nearly similar frequencies exist with sufficient amplitudes, providing the opportunity for cross-talk.

Thus radiation is an electromagnetic field that contains a spectrum of all the frequencies of oscillations on the surface of all matter within direct view. These frequencies (colors) do not change over distance, even galactic distance, except for very small **Doppler effects** and **gravitational redshift**, which are not important for our discussion. The amplitude (brightness) at each frequency, on the other hand, decreases in inverse proportion to the propagation distance as the signal spreads out over the surface of an expanding hemisphere due to its radiating from a point on a surface. We talk about rays of light; we think in terms of rays of light, but all that is needed is line of sight. There is nothing physical "occupying" that line of sight.

This is how we see. Every molecule out there, making up objects that we can see, is oscillating at some frequency in the visible light spectrum between approximately 380 terahertz (red) and 790 terahertz (violet). Note in Figure 4.3, that **visible radiation** (light) has wavelengths (sizes) in the range of 0.5×10^{-6} meters, comparable to the size of molecules in the visible world and the size of cells in our eyes (Figure 4.3). Each frequency in this range is a unique shade of some color in the rainbow. These frequencies cause molecules in the cells in the rods and cones of our eyes to resonate. More specifically, the retina in our eyes contains three types of cones, normally labeled L, M, and S, that resonate respectively with red, green, and blue frequencies, and our brains convert the relative signal levels from these cones to the appropriate shades of color, much as an RGB (red-green-blue) computer monitor converts red, green, and blue digital signals to the spectrum of colors on the computer screen.

Note, in Figure 4.3, that there is a temperature associated with each frequency. It is possible that the cones in our eyes respond to the temperature of light, not the frequency, but this is a matter of semantics. When a cone absorbs some frequency of oscillation, the frequency stimulates a chemical reaction in the photoreceptor perhaps by raising the cell to that color temperature but only for an instant. In other words, at the molecular level, frequency and temperature approach the same thing. Cells resonating to the "hottest" frequencies can become damaged if their amplitudes (brightnesses) are high enough. The eyes of young children have more sensitivity to ultraviolet radiation than the eyes of adults, because they have not yet been damaged by too much sunlight. Similarly, eyes in bottom-feeding fish are very sensitive to ultraviolet light that penetrates tens of meters into the ocean but is not bright enough to damage the fishes' photoreceptor cells.

Thus, radiation is a self-propagating electromagnetic field induced by the motion of charge in microscopic oscillators located on the surface of a radiating body. Radiation typically consists of a broad spectrum of frequencies (a.k.a. "colors"), each with a specific amplitude ("brightness"), resulting from the temperature-dependent frequency and amplitude of oscillation of each of the bonds that hold matter together. Radiation from several different sources coexists within the electromagnetic field without interacting until the field comes in contact with matter.

What Is Temperature?

In 1900, Max Planck figured out, by trial and error, that by adding a -1 in the denominator of an equation (Figure 4.5) developed earlier by Wilhelm Wien,[94] he could make the

$$B_v(T) = hv \left(\frac{2v^2}{c^2} \right) \left(\frac{1}{e^{hv/k_B T} - 1} \right)$$

Figure 4.5 Planck's law for spectral radiance (B_v) as a function of temperature (T) in degrees Kelvin, Planck's constant (h), frequency (v), the speed of light (c), and the Boltzman constant (k_B).

equation calculate spectral radiance that matched observations not only at low frequency, but also at high frequency. This equation applies to a "black body" at thermal equilibrium. A black body is simply a perfect absorber and emitter of radiation—it absorbs and radiates all frequencies present. It is the color black because none of the incoming radiation is reflected, scattered, or radiated back from it. A non-black body can be thought of as absorbing all radiation except its particular color (for example, green), which it, in effect, reflects, scatters, or radiates back. Thermal equilibrium means that thermal energy (heat) is no longer flowing from one point in the body to another point, except when there is an internal source of heat from which heat moves uniformly outward toward the radiating surface to replace the energy that is being radiated away from it.

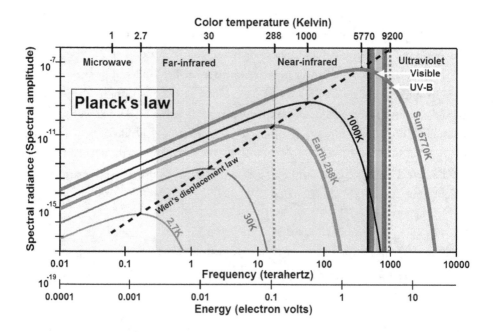

Figure 4.6 Planck's law shows that radiation from a warmer body has higher spectral amplitudes of oscillation at all frequencies than does radiation from a cooler body and exhibits its greatest amplitude at a higher frequency, following Wien's displacement law (black dashed line). Each solid line shows the spectral amplitude radiated from a body at thermal equilibrium for the temperature shown. Note that solar ultraviolet-B (UV-B) radiation (purple dotted line) has much higher energy and much hotter temperature on these log scales than infrared radiation from Earth (green dotted line).

Given these conditions, temperature can be defined "as that which is equal [i.e., uniform or homogeneous] when heat ceases to flow between systems in thermal contact."[95] Differences in temperature provide the thermodynamic "slope" down which thermal energy (heat) can flow. Since thermal energy flows via resonance, it can only physically flow from higher amplitude of oscillation to lower amplitude of oscillation at a given frequency. We all know from personal experience that heat only "flows" spontaneously via **conduction** in matter from higher temperature to lower temperature, i.e., from a heat source to a heat sink. We also observe that matter radiates heat only into colder air and that radiation must be "hotter" than a body of matter in order to be absorbed by the matter. You would not expect to get warm by standing next to a cold stove. This universally valid observation that heat flows from higher temperature to lower temperature is known as the second law of thermodynamics, and there exists no known credible instance of it ever having been violated.

Each solid line in Figure 4.6 shows the spectrum of frequencies and the amplitude at each frequency radiated from a body of matter at a specific absolute temperature in degrees Kelvin (K). At zero Kelvin, the bonds holding the matter together are too short, and too cold, to oscillate. It is physically impossible to cool a body colder than absolute zero Kelvin. Zero Kelvin equals -459.67 degrees Fahrenheit (-273.16 degrees Centigrade).

To review the most essential feature of Figure 4.6, each solid line represents a spectrum of frequencies of oscillation, and the amplitude of oscillation, at each frequency that makes up the radiation from a body of matter at a specific absolute temperature in degrees Kelvin (K). These are the same frequencies and amplitudes with which all the submicroscopic atomic oscillators on the surface of the radiating matter that induced the radiation are vibrating. The Planck curves define what the amplitude of oscillation is at a specific frequency for a body at some specified temperature.

Note that the lines do not intersect—they do not cross each other. This is extremely important because it indicates that when you increase the temperature of a body of matter, you increase the amplitude of oscillation at every single frequency and you also increase the frequency of the maximum amplitude of oscillation, as shown by the dashed black line, the equation for which is known as Wien's displacement law.

While the amplitudes at all frequencies decrease proportionally as the distance that the radiation travels, the frequencies (colors) do not change, and neither does the shape of the Planck curve. Thus, it is useful to characterize radiation by the frequency (color) that has the maximum amplitude of oscillation, i.e., the color temperature shown in Table 4.1 and on the x-axis at the top of Figure 4.6. Radiation does not possess a temperature, because radiation is not matter, and only matter (solid, liquid, or gas) can have temperature. It does, however, carry the spectrum of frequencies that cause temperature in matter. Color temperature is the temperature of a radiating body that emits a Planck spectrum of frequencies (Figure 4.6). It is the maximum temperature to which this radiation can warm an absorbing body of matter, given long enough exposure.

Applying these principles to greenhouse warming theory presents a fundamental problem. Greenhouse warming theory assumes that infrared radiation from Earth can be absorbed by greenhouse gases in the atmosphere, which then radiate infrared energy back to Earth's surface, where it is reabsorbed, warming Earth. This, however, is physically impossible because Earth's radiation can only make a body of matter as warm as Earth is, nothing warmer, and therefore it cannot heat Earth. If I surround a non-living body of matter with a perfect reflector that sends all the infrared energy radiated back to the matter, the matter would not get warmer. This is why a non-living thermal body cannot warm itself. This will be discussed in more detail in Chapter 10.

In summary, temperature is the macroscopic expression of a broad spectrum of microscopic oscillations of the bonds that hold matter together, described by the empirically formulated Planck equation shown in Figure 4.5, now known as **Planck's Law**. The curves in Figure 4.6 show the "natural" amplitude of oscillation at each frequency for a body at any of the specified temperatures. If you specify the temperature at the surface of a black body, then from Planck's law in Figure 4.5, you can determine the amplitude of oscillation at each frequency under normal ("natural") circumstances.

Why Is Thermal Energy Equal to Frequency Times a Constant?

What is thermal energy? In 1900, in order to derive the empirical equation shown in Figure 4.5, Max Planck needed to postulate that the energy (E) of a single atomic oscillator is simply equal to its frequency (ν, the Greek letter nu) times a constant (h), giving the expression $E=h\nu$. This constant of proportionality (h), now known as the **Planck constant**, is simply the amount of energy contained in an oscillation of one cycle per second (4.135667517 x 10^{-15} electronvolts (eV) per cycle per second). $E=h\nu$ is known today as the **Planck-Einstein relation**,[96] which is mistakenly thought of as the energy of a photon, as discussed below.

Note that the exponent in the denominator of Planck's empirical equation in Fig. 4.5 is equal to $h\nu/k_bT$, the energy per atomic oscillator at a given frequency ($h\nu$) divided by the energy at a specified temperature (T) in degrees Kelvin (k_bT). k_b is the Boltzmann constant, the amount of energy per degree Kelvin, commonly thought of as the bridge between macroscopic and microscopic physics. This ratio ($h\nu/k_bT$) was first written by Wilhelm Wien in 1896 in the Wien approximation,[97] which describes high-frequency radiation quite well. As noted, it was Wien's approximation that Planck modified slightly so that it would work at all frequencies.

A Problem With Spectral Radiance As Used in Planck's Law

According to Planck's empirical formula, spectral radiance is defined on the y-axis as the number of watts (joules of energy per second) at each frequency that passes through a square meter of surface area perpendicular to the direction from the emitting source per unit solid angle.[98] This exercise in empirical curve fitting, however, introduces a serious problem, of which Planck himself appears to have been unaware. As formulated, the formula assumes, in contradiction of the Planck-Einstein relation E=hν, that energy is the same at every frequency because energy is plotted on the y-axis, not the x-axis, that the amount of energy at a given frequency is a function of its amplitude, and that the total amount of energy is equal to the area under the curve. In reality, however, the empirically fitted curve does not physically describe the actual energy distribution in the spectrum. Energy does not remain constant across the spectral bandwidth; it increases to the right with increasing frequency, as the Planck-Einstein relation states (E=hν) and as labeled on the lowest horizontal axis in Figure 4.6. Energy is a function of frequency only. It is not a function of amount and is not related to the area under the curve. As Feynman stated, energy does not occur "*in little blobs of a definite amount.*"

Unfortunately, all climate models assume that available energy is a function of amount and of the area under the curve in Figure 4.6, as a result of which energy calculations in the infrared portion of the spectrum have been greatly overvalued while those in the ultraviolet portion have been greatly undervalued (Figure 1.10 on page 16). It is not surprising, therefore, that climate scientists have had such difficulty in balancing Earth's energy budget.[99]

These curves of Planck's law were measured, and are still measured today, by determining how much the temperature of a small body of matter, typically contained in a thermocouple, thermopile, or resistor, is raised when illuminated by a narrow bandwidth of radiation. Thus, we are measuring the thermal effect of radiation, not an actual physical property of radiation. This subtle point is very important.

Electromagnetic radiation, or more narrowly, visible light, with which we are used to dealing, has two observable physical properties: color (frequency of oscillation) and brightness (amplitude of oscillation). A more realistic picture of electromagnetic energy distribution than Planck's law (Figure 4.6) would show microscopic amplitude of oscillation (brightness or amplitude) on the y-axis as a function of the microscopic frequency of oscillation (energy) on the x-axis. In the Planck curves, temperature in the macroscopic world is related to these microscopic oscillations, which were plotted on the basis of the conduction of thermal energy within matter leading to thermal equilibrium. Measuring microscopic amplitude of oscillation would have been very difficult in 1900, and it is still much more difficult than measuring thermal effect. In order to arrive at a clear understanding of the physics of microscopic energy and macroscopic temperature, however, these measurements need to be made. In anticipation of this important step being taken, I will refer to the Planck's law curves as if they plotted microscopic amplitude as a function of microscopic frequency, fully recognizing that the precise shape of the curves and the scale on the Y-axis may vary to some extent when measured according to the aforementioned criterion.

Thermal energy is frequency. The usual ("natural") amplitude of oscillation at each frequency is specified by Planck's law as a function of absolute temperature. Your heart beats, while resting, at a frequency normally around 72 cycles (beats) per minute. Energy is the heartbeat of everything in the Universe, the 'go' of the Universe, but it beats at rates of trillions of cycles per second. To make viable climate models, the "go" must be properly included.

Energy Is Clearly Observed to Increase With Increasing Frequency of Oscillation

This spectrum of increasing energy with increasing frequency makes observational sense when we look at the list of chemical effects as a function of frequency in Table 4.1 and the overall electromagnetic spectrum shown in Figure 4.3. It is well known that very high frequency gamma rays (nuclear radiation, right side of Figure 4.3) have enough energy to be lethal if absorbed in very small quantities. Lower frequency X-rays have enough energy to destroy cancer cells when focused at high dosages, but can be used at very low dosages to prevent damage when imaging the interiors of human bodies. It is common knowledge that lower frequency solar ultraviolet radiation has enough energy to burn human skin, that still lower frequency visible light has enough energy to power photosynthesis, and that much lower frequency infrared radiation does not have enough thermal (chemical) energy to do either, but that it can provide comforting warmth on cold days. Note that dosage is proportional to energy times the length of time of the exposure, as is typical of dynamic processes.

Clearly, then, the effectiveness of electromagnetic radiation in bringing about certain changes in matter is governed by a threshold effect. Radiation having a frequency that is higher than the threshold (E_{max} in Figure 4.4) will produce the change, whereas radiation having a lower frequency will not. A good example of this is the photoelectric effect, in which certain metals emit electrons when light shines on them. Maxwell's wave theory of light predicts that the more intense the incident light, the greater will be the energy with which the electrons should be emitted, but this is not what is observed.[100] The electrons are dislodged only when the light exceeds some frequency threshold typically in the blue to ultraviolet range. Light below this frequency cannot cause emission of electrons no matter the amount (intensity) or length of exposure. Above the threshold frequency, the rate at which electrons are ejected is directly proportional to the intensity of the light source, but there is no effect on the kinetic energy of individual electrons. The photoelectric effect is dependent only on frequency, as shown by E_{max} in Figure 4.4,[101] not on amplitude, intensity, or amount.

In **photochemistry**, the Planck-Einstein relation, $E=h\nu$, is widely used in chemical equations to signify the minimum radiant energy (frequency) required at the microscopic level to cause a chemical reaction, such as the **photodissociation** or **photoionization** of oxygen, to take place. It is appropriate to think of the high frequency (right-hand end) of the Planck curves in Figure 4.6 as being the chemical front, the frequencies where chemical reactions will first take place. We will learn in the next chapter that this explains how solar energy is absorbed within Earth's atmosphere.

In summary, firstly, radiant chemical energy is clearly observed to be solely a function of frequency as stated in the Planck-Einstein relation, and not a function of amount, amplitude, intensity, or bandwidth. Some minimum frequency determines whether there is enough energy for a chemical reaction to take place. The amount, brightness, intensity, or amplitude of incident radiation does, however, determine the *rate* at which these microscopic reactions can take place, depending on how much amplitude of oscillation is transferred by resonance. We are not burned by distant starlight, for example, because its amplitude is so low.

Secondly, from Planck's curves shown in Figure 4.6, it is clear that as the temperature of the radiating body increases, the amplitude of oscillation at every frequency increases and the frequency of the highest amplitude of oscillation also increases. Heat flows at each frequency from higher amplitude of oscillation to lower amplitude of oscillation. Therefore radiation from a cooler body cannot warm a warmer body—the amplitudes are not high enough. Furthermore, radiation from a given body cannot warm the same body no matter how efficiently the radiation is reflected back—the radiation does not have higher amplitude at each frequency. The radiation, therefore, cannot increase the amplitudes of oscillation within the body as the radiation is received, and therefore it cannot increase the temperature of the receiving body.

Thirdly, and most importantly, radiant chemical energy throughout the electromagnetic spectrum is observed to increase with frequency (Figures 4.3 and 4.6). $E=h\nu$ is the energy of an atomic oscillator, and it is this energy that is conducted through matter, radiated from the surface of matter, and absorbed by other bodies of matter. Radiant chemical energy is not a function of amplitude, amount, intensity, or bandwidth, as assumed by most physicists and climatologists. This is the primary reason why greenhouse warming theory and climate models do not appear to be correct.

This is likely a shocking conclusion for anyone trained in physics. It is not the way we have traditionally thought about energy. Since Maxwell wrote his influential equations in the 1860s, nearly all physicists have assumed that electromagnetic radiation travels as waves through space, that energy is proportional to the square of the amplitude of these waves and to their bandwidth, and that energy comes "*in little blobs of a definite amount*" that can be added together. Nevertheless, after Einstein's paper on the photoelectric effect in 1905, the concept that light travels as particles or as wave-particle duality began to get some traction.

Why Can't Thermal Energy Propagate as Waves?

The foregoing discussion of electromagnetic radiation is completely described in terms of frequency and amplitude of microscopic oscillations of the bonds that hold matter together. There is no mention of wavelength. Physically, it is not clear how wavelength could be involved in absorption, conduction, emission, or resonance. All the same, light, when interacting with matter, displays properties, such as reflection, refraction, dispersion, diffraction, **birefringence**, and interference, that we traditionally explain using the mathematics of wave forms. Also, scientists are accustomed to calculating wavelength by dividing the velocity of light by frequency, but that is mathematics. Is it physics? We tend to think about electromagnetic waves based on our experience with mechanical waves in matter, such as water waves, sound waves, and compressional, shear, and surface seismic waves. Mechanical waves, however, have very different physical properties from electromagnetic radiation.

A mechanical wave, whether longitudinal (compressional, oscillating in the direction of travel) or transverse (shear, oscillating perpendicular to the direction of travel), is a physical disturbance that propagates through solid matter. Compressional waves can also propagate through a liquid or a gas. Each molecule is displaced by the wave's kinetic energy and is then normally restored to its original position by the bonds or pressure that hold the matter together. The energy of the wave is the kinetic energy required to deform the matter, allowing the wave to propagate, and it is typically proportional to the square of the amount of displacement (wave amplitude). The wave works against the bonds or pressure holding the matter together, and it is therefore attenuated (decreased) with distance. The stiffer the material, the more work is necessary to propagate the wave, and the faster the attenuation. High frequencies are typically attenuated more rapidly than low frequencies.

We feel, and/or observe with our eyes and instruments, seismic and water waves as they deform matter with frequencies most commonly in the range of 0.005 to 100 cycles per second, and we hear sound waves with frequencies primarily in the range of 20 to 20,000 cycles per second. Transverse waves can only propagate through elastic matter, whereas longitudinal waves can propagate through all types of matter, including liquids and gases. Neither can propagate through space. The motion of each molecule in matter is connected by its bonds or by pressure to the motion of all surrounding molecules, so that waves from different sources are observed to interact and interfere.

This physical reality that everything is interconnected is expressed mathematically as the plus signs in a Fourier series. A Fourier series approximates a specific waveform at a specific location as the sum of a large, if not infinite, series of terms, each consisting of an amplitude times the sine and/or cosine of a different wavelength or frequency. Thus, any wave can be described as an infinite series of a little bit of each component frequency all summed together. It is the bonds or pressure holding solid matter together that provide the physical basis for the plus signs, or "addability," in the Fourier series. Everything involved with mechanical waves is interconnected.

We observe that the physical properties of electromagnetic radiation (light), however, are distinctly different from the physical properties of mechanical waves. Light radiated by a black body at thermal equilibrium contains a broad spectrum of frequencies (colors) with different amplitudes (brightnesses) at each frequency, as described by Planck's law shown in Figures 4.5 and 4.6. As I have pointed out, the frequencies and amplitudes of light originate from the frequencies and amplitudes of the atomic oscillators on the surface of the radiating body. I have also noted that in air and space, light frequencies are observed to exist totally independently, maintaining their individual integrities. They do not interact, except in the immediate presence of solid matter, and they do not change over distance, even galactic distance. Electromagnetic radiation in air and space can be thought of as a Fourier series without plus signs. There is some red light, some blue light, some yellow light, etc., that do not interact until the full spectrum interacts with matter, as in a rainbow or prism (Figure 4.7).

Amplitude (brightness), on the other hand, decreases in inverse proportion to the propagation distance as the signal spreads out over the surface of an expanding hemisphere due to its radiating from a point on a surface. There is no evidence of attenuation (reduction) of amplitude by friction.

The co-existence in electromagnetic radiation of numerous frequencies that do not interact is shown quite clearly, as mentioned above, by radio signals transmitted and received at millions of very precise frequencies. These signals do not interfere with each

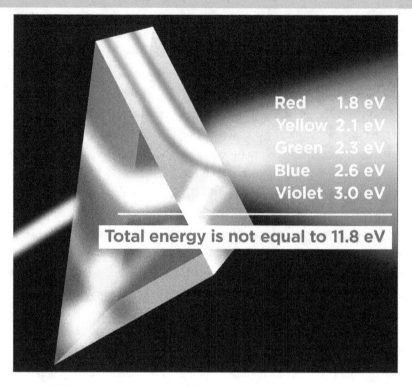

Red	1.8 eV
Yellow	2.1 eV
Green	2.3 eV
Blue	2.6 eV
Violet	3.0 eV

Total energy is not equal to 11.8 eV

Figure 4.7 White light entering from the left is spatially separated by a prism into its component colors. The total energy contained in white light is not the sum of the energies of the colors (frequencies) that it contains. White light contains visible energies ranging from 1.65 to 3.26 electronvolts (eV) (Table 1). Summing the component energies yields 11.8 electronvolts, an energy level that is characteristic of ultraviolet-C radiation, which is well outside the actual distribution of energies. Computer models that calculate the energy absorbed by greenhouse gases mistakenly integrate energy across all frequencies (wavelengths) involved, leading to erroneous results.

other except when receivers are too close to two or more transmitters putting out the same, or very similar, central frequencies. Similarly, colors that we observe do not interact with each other in air and space. Otherwise, everything we see would be blurry.

Why Can't Thermal Energy Propagate as Photons?

If we think of a photon as a packet of energy that interacts with a molecule of gas, then light cannot physically travel as photons. When a molecule of a greenhouse gas absorbs electromagnetic radiation, it only absorbs by resonance very narrow spectral lines of energy, as shown in Figure 4.8 for the primary band of energy absorbed by **carbon dioxide** around

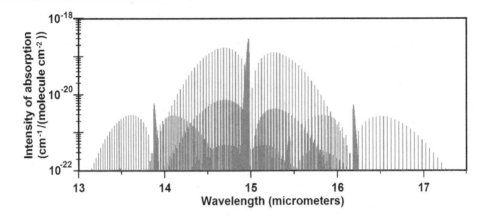

Figure 4.8 The spectral lines of energy (red) absorbed by a molecule of carbon dioxide are the resonant frequencies of the normal modes of oscillation of all the degrees of freedom of all the bonds that hold the molecule together. The energy absorbed is determined by the structure of the carbon dioxide molecule at the location of the molecule. It is traditional to plot intensity as a function of wavelength, the velocity of light divided by the frequency, even though waves may not be involved.

the 14.9 micrometer band shown in Figure 1.10 on page 16. This band is shown, as is traditional, as a function of wavelength, which we think of as the velocity of light divided by frequency. These spectral lines have been well studied in the laboratory and tabulated[102] in exquisite detail by spectral physicists for all kinds of gases because they are used from near at hand to far distant galaxies to determine the chemical content of gases. Each spectral line is a resonant frequency of some degree of freedom of one of the bonds involved. Thus, the details of the packet of energy absorbed, something we think of as a photon, are determined by the physical nature of the absorbing molecule at the location of the absorbing molecule. Such "made-to-order" photons cannot travel from Sun or other hotter bodies. If they did, there would have to be a different type of photon for each spectral line of each different type of molecule. Furthermore, light is continuous spatially at any scale. Light spreads with distance travelled. How, then, could photons be spaced close enough when they leave a star in a far distant galaxy in order to appear continuous from Earth?

Potential Radiant Temperature and the Amplitude of Oscillation

Showing that energy in radiation is a function of frequency, not amplitude, leads to a crisis in how we describe the amount of electromagnetic radiation (light) that we receive. Existing descriptions and units treat amount of radiation in terms of power per unit area—

such as watts per square meter, which is joules per second per square meter. These units are included in the definitions of radiance, radiosity, radiant exitance, luminance, and intensity and to some extent in the more subjective definitions of brightness and lightness.

Radiation, as it leaves the surface of a piece of matter that is at thermal equilibrium, has a distribution of frequencies of oscillation and amplitudes of oscillation as a function of the temperature of the matter, as described by Planck's law (Figures 4.5 and 4.6), corrected for the reality that spectral radiance on the y-axis should be amplitude of oscillation, as discussed above. Radiation cannot have a temperature because it does not have mass, but we can think of radiation as having a potential radiant temperature, meaning that if it were absorbed in sufficient quantity by a colder body of matter, the radiation would have the potential to warm that body to the same temperature as the emitting body—it contains the frequencies and amplitudes that have the potential to raise the new body to the color temperature of the emitting body, but no warmer.

That potential can only be realized at very close range, however. As the radiation travels away from the surface of the emitting matter, the frequencies all stay the same, but all the amplitudes are observed to decrease as a function of distance. This means, as shown in Figure 4.6, that the potential radiant temperature of the radiation decreases with distance. Solar radiation near Sun has a potential radiant temperature of 5770K, but by the time it reaches the Earth/atmosphere system, it has a potential radiant temperature closer to 255K (-0.4°F or -18°C), the temperature at which most physicists think a planet the size of Earth without an atmosphere and at a distance of one astronomical unit from Sun (i.e., the mean Earth-Sun distance) should be. This is essentially equal to the temperature of the **stratopause** whose estimated average temperature is thought to be around 5°F (-15°C, 258K) as discussed above.[103] It appears from Figure 4.6 that in order to preserve the form of the Planck's law curves, the amplitudes of the peak frequencies must decrease faster than the amplitudes at lower and higher frequencies, but this probably reflects how the energy is absorbed by the absorbing matter, not how the spectral nature of radiation changes with distance. It is clear, as explained in the next chapter, that the highest frequencies, which have the highest chemical energy of oscillation, are absorbed first as solar radiation penetrates the atmosphere.

Referring back to the properties of an atomic oscillator shown in Figure 4.4 and Planck's law plotted in Figure 4.6, the amplitude of oscillation at the microscopic level of the oscillator has a very specific "natural" or "normal" value based on the temperature of the radiating body and the frequency of oscillation so that the thermal energy of oscillation is specified as $E=hv$, the Planck-Einstein relation. As the radiation propagates, however, this "normal" amplitude, representing the thermal energy of oscillation, decreases without altering the frequency, which represents the chemical energy of oscillation, an invariant quantity.

Monochromatic lasers add energy at a specific frequency in order to increase this amplitude of oscillation—the thermal energy—without changing the frequency—the chemical energy. As the amplitude of oscillation is increased, the thermal energy of oscillation approaches the energy threshold (E_{max}) causing the molecule to come apart, i.e. to "melt or vaporize." Thus, a red laser can do much more damage to matter than red light, and a higher frequency green laser can do even more damage than a red laser.

A laser also forms light in such a way that its beam does not spread, and, therefore, the amplitude does not decrease with distance. Small, handheld, green laser pointers have become a major hazard when pointed perversely from the ground toward airplanes because they can temporarily blind pilots.

We observe that light has color and brightness. The 6 million cone cells in the human retina are split into three groups that contain pigments making them most sensitive to red light (around 532 THz), green light (around 561 THz), and blue light (around 714 THz). "To determine intensity, the visual system computes how many photoreceptors are responding."[104] The brain processes all these inputs to construct the color images we see.

The mathematics of this new way of looking at radiation will have to be worked out experimentally, but meanwhile we do not have a word for increased "brightness" or "intensity" of light because those words have special meaning related to an old way of looking at radiation.

Thermal Energy Is a Spectrum of Frequencies

This book explains that internal thermal energy and radiant thermal energy constitute a spectrum of frequencies that exists with a certain amplitude of oscillation at each frequency, rather than being "*little blobs of a definite amount.*" Physicists I talk to about this have trouble thinking of frequency as a "thing." I had the same problem. It is not the way we have been trained to think. It is much easier to think of wavelength as a thing, as wavelength is something we can visualize. After all, natural philosophers and scientists have debated for 2400 years whether light—something we cannot see until it interacts with matter—travels through space as either a wave or a particle—things we can see. There is a logical disconnect here. Electromagnetic radiation (e.g., ultraviolet and visible light and thermal radiation) comprises a broad band of frequencies; the energy at each frequency is equal to the frequency times the Planck constant; and the frequencies form a continuum, i.e., they are not discrete, except when absorbed by normal modes of oscillation of a single molecule of gas, yet they do not interact or interfere in air or space. Visualizing this continuum is not easy, and visualizing how amplitude spreads physically over the surface of an expanding sphere as the energy propagates is not easy, either, but both are clearly happening. Maybe that is why understanding energy in radiation has eluded scientists for so long.

Understanding that radiant energy is equal to frequency times a constant and that the highest energy, highest frequency radiation along the chemical activity front on the high-energy end of the Planck curves interacts first with atmospheric gases are the keys to understanding the structure of Earth's atmosphere and how the atmosphere protects life on Earth from the "hottest," most photochemically destructive solar radiation—concepts that are described in the next chapter.

CHAPTER 5

HOW DOES THE ATMOSPHERE PROTECT EARTH FROM SUN'S "HOTTEST" RADIATION?

"In fact, the thickness of the Earth's atmosphere, compared with the size of the Earth, is in about the same ratio as the thickness of a coat of shellac on a schoolroom globe is to the diameter of the globe. That's the air that nurtures us and almost all other life on Earth, [and] that protects us from deadly ultraviolet light from the sun."
—Carl Sagan, 1995

T heoretically, a planet without an atmosphere at Earth's distance from Sun should have a mean surface **temperature** of around -0.4°F (-18°C), but Earth's mean surface temperature is closer to +59°F (15°C). Therefore, the presence of the atmosphere warms Earth approximately 59°F (33°C). Many climatologists claim that it is the presence of **greenhouse gases** in the atmosphere that explain this warming. Yet the reality discussed in this chapter is that heat above the **tropopause**, at an average altitude of 10.6 mi (17 km), comes primarily from Sun, not from Earth.

Earth's atmosphere has a complicated temperature structure that varies primarily with latitude and with time of day. The typical values at mid-latitudes are shown in Figure 5.1, based on the U.S. Standard Atmosphere of 1976.[105] Temperature in the **troposphere**, the lowest part of the atmosphere, decreases with increasing altitude at a lapse rate of 5.5°F per thousand feet (10°C per km) if the air is very dry, or 3.0°F per thousand feet (5.5°C per km) if the air is very moist. The average lapse rate is well observed to be around 3.5°F per thousand feet (6.4°C per km).

Earth's surface is heated primarily by absorbing radiant energy from Sun during the day. The Sun-warmed surface heats the air above, much as the burner on an electric stove heats the air above it. The heated air is less dense and therefore rises, cooling as it expands with increasing altitude and decreasing pressure—a process called convection. The troposphere is heated primarily from below in this manner. While Earth's surface also loses heat by **radiation**, the primary flow of heat through the troposphere is by convection, especially during storms.

The greatest convectional heating of the troposphere is in the tropics, and the least convectional heating is in the polar regions. This difference in heating within the troposphere creates a temperature gradient that drives heat poleward, creating the weather systems and surface ocean currents that influence our daily lives.

All this changes at the tropopause—the top of the troposphere—however, where temperatures stop decreasing with increasing altitude. The tropopause is at an average altitude of 10.6 mi (17 km), but it varies from 4.3 mi (7 km) in polar regions during winter to 12.4 mi (20 km) in the tropics year round. The troposphere contains approximately 76% of the atmosphere's mass and 99% of its water vapor. Turbulence, caused primarily by differences in temperature along Earth's surface, is abruptly and significantly reduced at the tropopause. Above this important boundary, the air becomes stratified, forming the **stratosphere**, because all of the atmosphere above the tropopause is heated primarily by radiant energy directly from Sun rather than by Earth's heat convected upward from below. The mechanism by which this heating takes place is a process called **photodissociation**, which works in the following way.

When a molecule of gas absorbs radiant energy of a sufficiently high frequency to break the bonds holding the atoms together (E_{max} in Figure 4.4 on page 51), the ions or atoms fly apart at very high velocities. In a closely related process called **photoionization**, one or more electrons are lost or gained by the products, thus converting them into ions. In photodissociation, all of the **internal energy** stored in the affected bonds is converted directly to **kinetic** energy of the ions or atoms. Kinetic energy is simply one-half the mass of the atom times the atom's velocity squared. Since the temperature of a gas is proportional

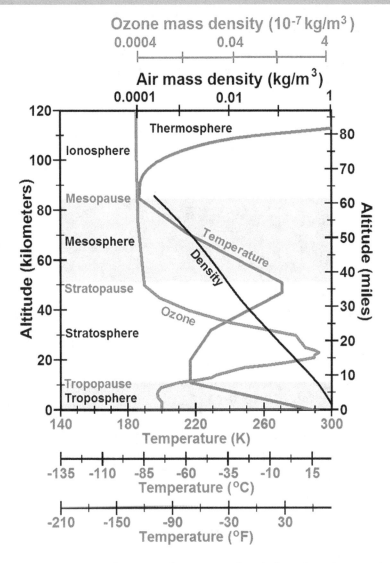

Figure 5.1 Temperature decreases with increasing altitude throughout the troposphere, which is warmed primarily by convection of heat from Earth's surface. Temperature increases with altitude throughout the stratosphere, which is warmed primarily by solar-caused photodissociation of oxygen, ozone, and other species. Temperatures decrease with altitude throughout the mesosphere, the radiant region of the Earth/ atmosphere system losing heat to space. Temperatures warm with increasing altitude above the mesopause, warmed primarily by solar-caused photoionization of nitrogen, oxygen, and other species, but here densities are so low that gas temperature is no longer meaningful and measured temperatures are actually quite low.

to the average kinetic energy of all the molecules, atoms, and ions making up the gas, photodissociation and photoionization are the most efficient ways to convert internal energy stored in chemical bonds to temperature of the gas. This is how the atmosphere above the tropopause is heated.

How Does the Upper Atmosphere Absorb the Highest Energy Solar Radiation?

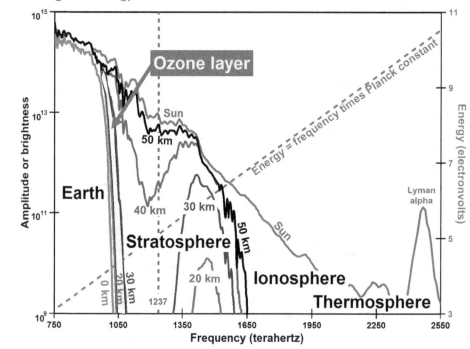

Figure 5.2 The highest frequency, highest energy, solar radiation is absorbed high up in the atmosphere. The red line shows the amplitude of solar radiant energy received at the top of Earth's atmosphere. Almost all radiation with frequencies greater than 1650 terahertz is absorbed above 50 km (31 mi) (black line). Much radiation with frequencies above 1090 THz is absorbed above the ozone layer (yellow).

Figure 5.2 shows the amplitude of the solar radiation that reaches the top of Earth's atmosphere (red line) and the amplitude of the solar radiation that reaches an altitude of 34 mi (50 km) (black line), as a function of frequency.[106] Note that all solar radiation with frequencies greater than approximately 1650 **terahertz** (THz) has been absorbed above the top of the stratosphere, known officially as the **stratopause**, ranging from altitudes of 31 to

34 mi (50 to 55 km) (labeled in Figure 5.1). Temperatures fall throughout the **mesosphere**, from above the stratopause to the bottom of the **mesopause**, beginning around 53 mi (85 km), and then more slowly to the top of the mesopause, also known at the **turbopause**, around 62 mi (100 km). Convection, turbulent mixing, plays a major role throughout the mesosphere just as it does in the troposphere, where temperatures also decrease with increasing altitude.

Above the turbopause, molecular diffusion dominates, meaning that the chemical composition of the atmosphere varies as a function of chemical species and chemical species vary as a function of the highest solar radiation to penetrate to a given altitude. Above 37 mi (60 km), solar energy is high enough in the ultraviolet-C (UV-C) and in the extreme ultraviolet, with energies in excess of 6.82 electron volts (Table 4.1), to ionize primarily nitrogen (N_2) and oxygen (O_2), which together make up 99.03% of the atmosphere on average. This is where the **ionosphere** begins, and from here, it extends upwards to around 620 mi (1000 km) above Earth, depending on solar activity.

The **thermosphere**, which coincides through most of its altitude range with the ionosphere, extends from the mesopause, around 53 mi (85 km) above Earth, to altitudes somewhere between 310 and 620 mi (500 to 1000 km), also depending on how much high-energy radiation is arriving from Sun. Temperatures are, therefore, also highly dependent on solar activity and can rise to 3630°F (2000°C), but the air at these altitudes is so thin that there are not enough atoms to transfer much heat, so that a thermometer would register significantly below 0°C. Recall that, as shown in Figure 4.3 on page 48, the higher the frequency, the higher the energy, and the higher the temperature. Temperatures are greatest high in the thermosphere where the most energetic solar radiation is being absorbed, typically leading to photoionization. The higher the frequency (energy) being absorbed, the higher the internal energy of the absorbing bonds, the higher the velocity of the separating ions, and the higher the "temperature."

The electrically charged ions interact with the lines of force of Earth's magnetic field, forming brilliant auroral displays under the right conditions.

How Does Oxygen Absorb Solar Radiation?

The details of the absorption of solar radiation by oxygen are well understood by spectral physicists[107] and are summarized in Figure 5.3.[108]

All such absorptions occur at frequencies greater than 1237 THz (dashed blue lines in Figures 5.2 and 5.3). Photoionization of oxygen in the ionosphere is caused by frequencies greater than 2910 THz along the ionization continuum, in which

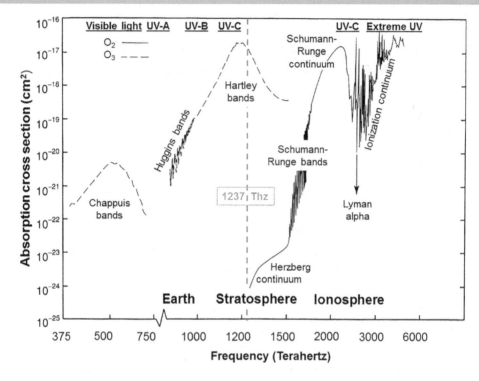

Figure 5.3 The amount of absorption of solar radiation by oxygen (solid lines) and by ozone (dashed lines) as a function of frequency is low in the Herzberg continuum, just to the right of the 1237 THz blue dashed line, explaining the minimal absorption by an altitude of 20 km between 1350 and 1600 THz shown in Figure 5.2.

photoionization occurs continuously throughout the frequency range. Photodissociation dominates absorption by oxygen in the thermosphere along the Schumann-Runge continuum. Pre-dissociation dominates absorption by oxygen in the mesosphere along the Schumann-Runge bands (spectral lines). Finally, the Herzberg continuum results from something known as a **forbidden transition** that dominates absorption by oxygen in the stratosphere. The inefficient absorption of frequencies between 1340 and 1600 THz explains why these frequencies have not been absorbed substantially by an altitude of 18.6 mi (30 km) as shown in Figure 5.2. In addition, many other trace gases are also dissociated in the stratosphere.

The bottom line, though, as shown in Figure 5.2, is that by an altitude of 30 km, which corresponds to the upper part of the **ozone layer**, nearly all solar ultraviolet-C energy (frequencies from 1033 to 2000 Thz) has been absorbed, causing dissociation of oxygen and other gas species and thereby warming the atmosphere from above.

How Does the Ozone Layer Protect Earth from Sun's "Hottest" Radiation?

The ozone layer, primarily 12 to 19 miles (20 to 30 km) above Earth's surface, is not a static layer of gas, but is a region of the atmosphere where sufficient solar **ultraviolet radiation** of frequency greater than or equal to 1237 THz is available to dissociate oxygen, leading to the formation of ozone, and sufficient ultraviolet radiation at lower frequencies is available to dissociate ozone. Ozone is dissociated most efficiently by solar ultraviolet-B radiation with frequencies around 967 THz, although some photodissociation is caused by frequencies as low as 887 THz.[109] Thus, ozone is created and destroyed continually in a sequence known as the Chapman cycle, which will be described in the next chapter. The ozone layer is the last chance to absorb solar ultraviolet radiation before it strikes Earth.

The green area in Figure 5.4 shows the frequency distribution of excess ultraviolet-A and ultraviolet-B radiation that reaches Earth when the ozone layer is depleted by 1%, as calculated by Sasha Madronich.[110] Actinic flux on the y-axis is defined as the total intensity of sunlight available to be absorbed by a molecule of air, including direct, scattered, and reflected radiation coming from all directions, as calculated by Madronich.[111] Actinic flux is a more detailed and complete representation of the "Amplitude or brightness" spectrum plotted in Figure 5.2. Note the substantial increase in UV-B and the lesser amount of increased UV-A radiation reaching Earth when ozone is depleted by only 1%.

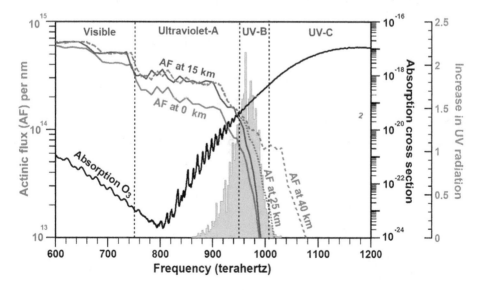

Figure 5.4 The green shaded area shows the frequency distribution of the increase in ultraviolet radiation reaching Earth when the ozone layer is depleted by 1%.

Recognizing that radiant energy is simply a function of frequency times the **Planck constant** (E=hv), and that the higher the frequency the higher the color temperature of the radiation, then it is easy to understand why the mean surface temperature of Earth is determined primarily by the highest frequencies of ultraviolet radiation reaching Earth as well as by the length of time Earth is exposed to this very "hot" radiation.

What Is the Primary Radiative Surface of the Earth System?

The major warming observed in Earth's atmosphere is in the stratosphere (Figure 5.1), where temperatures at the stratopause, 31 to 34 mi (50 to 55 km) above Earth's surface, are maintained at tens of degrees warmer than temperatures at the tropopause, at 5.6 to 10.6 mi (9 to 17 km) altitude. This warming is done primarily by solar ultraviolet energy dissociating molecular oxygen (O_2) and many other chemical species. The stratosphere acts as an "electric" blanket, in the sense that the energy to warm the blanket does not come from the body under the blanket, i.e. from Earth, but from another source, in this case, Sun.

Temperatures at the stratopause vary greatly with season but commonly range from 8.6 to 37°F (-13 to +3°C) in the tropics, -5.8 to 45°F (-21 to +7°C) in mid latitudes, and -4 to 68°F (-20 to +20°C) near the poles[112] and average close to 5°F (-15°C).[113]

Thus, solar energy causing photodissociation of oxygen and other chemical species in the stratosphere plays the major role in keeping Earth warmer than it would be in the absence of an atmosphere. The stratopause is the key radiant surface into space of the Earth/atmosphere system, and it can continue to radiate only because heat from both a Sun-warmed stratosphere and a Sun-warmed Earth rises continuously through the stratosphere below, replacing the heat radiated from the stratopause. Temperature drops with increasing altitude above the stratopause and below the mesopause just as in the troposphere, so that turbulence is likely active there.

The temperature profile shown in Figure 5.1 is the balance within the atmosphere of heat flowing into and out of the Earth/atmosphere system.

CHAPTER 6

HOW DO MINUTE AMOUNTS OF OZONE CONTROL CLIMATE?

"I can live with doubt and uncertainty. I think it's much more interesting to live not knowing than to have answers which might be wrong."
— **Richard P. Feynman**, 1981

O zone makes up only 0.3 **parts per million** (ppm) of the atmosphere, compared to 400 ppm for **carbon dioxide** and 209,460 ppm for oxygen. Carbon dioxide and oxygen are well-mixed throughout the atmosphere, while ozone is most prevalent in the **stratosphere**, with the highest concentrations of ozone approaching 10 ppm at altitudes of 17 to 23 mi (28 to 37 km) above Earth[114] (Figure 6.1a). The highest **partial pressure** of ozone (in millipascals, mPa) (Figure 6.1b) is at altitudes of 9 to 14 mi (15 to 30 km). Partial pressure of a gas is a measure of the thermodynamic activity of the gas molecules. Note how the lower, dark purple band on the right is essentially the **tropopause** from 5.6 mi (9 km) at the poles to 11 mi (17 km) in the tropics.

How can changes in such minute concentrations of ozone have the dominant effect on climate change? The short answer to this question is because ozone has a short lifetime

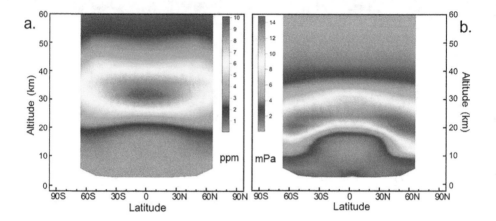

Figure 6.1 The greatest concentrations of ozone (ppm) (left) are at altitudes of 28 to 37 km (17 to 23 mi) above Earth. The greatest partial pressures of ozone (mPa) (right) are at altitudes of 15 to 30 km (9 to 14 mi). Colors show annual mean values from 1984 to 2004 measured by the Sage II satellite.

and is continually being created and destroyed by processes that, in both cases, turn solar **radiation** into stratospheric heat. Approximately 12% of the **ozone layer** is created every day.[115] This means that the average lifetime of a single molecule of ozone is only 8.3 days!

This continual creation and destruction of ozone absorbs 97 to 99%[116] of solar **ultraviolet radiation** with frequencies in the range of 950 to 1500 THz, comprising the Hartley absorption bands shown in Figure 5.3 on page 74. The results of absorbing this solar energy are:

1. splitting apart (dissociating) ozone molecules,
2. warming the lower stratosphere, and
3. protecting life on Earth from this DNA-damaging radiation.

When the amount of ozone in the ozone layer is reduced (depleted), more of this very "hot" radiation, primarily with frequencies between 900 and 1000 THz (green histogram in Figure 5.4 on page 75), reaches Earth, cooling the lower stratosphere, warming Earth, and increasing the risk of sunburn and skin cancer.

The ozone layer acts like a globe-circling heater, heating the lower stratosphere. It is powered by solar ultraviolet radiation, whose effect is greatest when Sun is directly overhead, and varying with latitude, time of day, and change of season. The **photodissociation** of oxygen and ozone within the layer provides and maintains the greatest warming effect anywhere in the Earth/atmosphere system. The continual cycle of dissociating oxygen to form ozone and dissociating ozone to form oxygen again is the primary photochemical process that maintains the **temperature** at the top of the stratosphere at tens of degrees warmer than the temperature at the base of the stratosphere (Figure 5.1 on page 71).

Neither the stratosphere nor life as we know it could exist on Earth without the ozone layer, yet as a product of the ozone-oxygen cycle, described in the following section, the ozone layer is the fragile Achilles heel of Earth's climate. We depend on it absolutely as we walk through life. By manufacturing large amounts of CFCs, we accidentally damaged the ozone layer. The damage has been very slowly improving since 1995, however, thanks to swift action via the **Montreal Protocol**.

What Is the Ozone-Oxygen Cycle?

In 1930, Sydney Chapman[117] outlined a cycle whereby ozone is continually created and destroyed, and it is known today as the Chapman Cycle. The ozone "layer" is not a layer of static gas. It is a dynamic region in the atmosphere where there is sufficient ultraviolet solar radiation and other physical conditions that promote the Chapman cycle. It is the

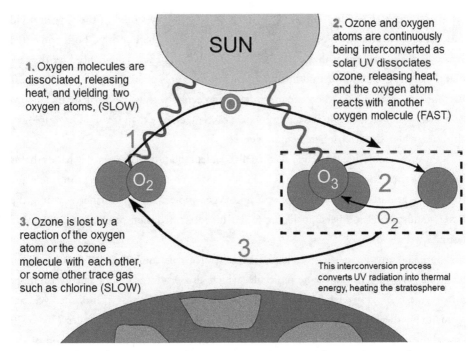

Figure 6.2 The Chapman cycle that continually forms and destroys the ozone layer 12 to 19 mi (20 to 30 km) above Earth.

Chapman cycle that gets degraded when humans release CFC gases into the atmosphere, causing ozone depletion.

Ozone (O_3) is created when an oxygen molecule (O_2) absorbs sufficient ultraviolet-C radiation with frequencies high enough to split (photodissociate) the oxygen molecule (O_2; see Figure 4.4 on page 51), forming two oxygen atoms (O) (see number 1 in Figure 6.2).[118]

This process proceeds at a relatively slow rate. Each oxygen atom (O) can then recombine quickly with an oxygen molecule (O_2) to form an ozone molecule (O_3).

The frequencies with the greatest absorption of ultraviolet-C radiation are mainly higher than 1249 THz, including the Schumann-Runge bands (1557 to 1703 THz) and the Schuman-Runge continuum (1703 to 2221 THz), but they also include the Herzberg continuum (1153 to 1249 THz), as shown in Figure 5.3 on page 74. Following their photodissociation, the two oxygen atoms (O) fly apart at very high velocity. As they collide with other gas molecules, these two atoms share their velocity with them, increasing the average velocity of all molecules in the gas (number 1 in Figure 6.2). The temperature of a gas is very sensitive to the velocity of its molecules because temperature is proportional to the square of the velocity, or, more specifically, is proportional to the average **kinetic** energy, of all gas molecules present. Kinetic energy equals one-half the molecule's mass times its velocity squared ($E_k = \frac{1}{2}mv^2$). In this way, the **internal energy** stored in the oxygen bond and the radiant energy absorbed to split (photodissociate) the bond are both converted very efficiently into atmospheric temperature. The heating effect is enhanced by the fact that the density of air in the middle of the ozone layer is only about 3% of the density near Earth's surface. Therefore, there are many fewer molecules with which to share the energy. As a result, each dissociation has a bigger effect on temperature than it would have at lower altitudes, where the atmosphere is more dense.

The ozone molecule (O_3) then absorbs ultraviolet radiation in the Hartley bands that is sufficiently energetic (1000 to 1500 THz, Figure 5.3) to photodissociate the ozone molecule (O_3) back into an oxygen atom (O) and an oxygen molecule (O_2), further warming the lower stratosphere (number 2 in Figure 6.2). This interconversion proceeds at a relatively fast rate.

Subsequently, an oxygen atom (O) so generated can combine with a new oxygen molecule (O_2) to form an ozone molecule, which also produces kinetic energy (number 3 in Figure 6.2), at a relatively slow rate. The net results of this oxygen-ozone cycle are to maintain the ozone layer predominantly at an altitude of 12 to 19 miles (20 to 30 km) above Earth and to maintain the temperature of the **stratopause**, at the top of the stratosphere, tens of degrees warmer than the temperature of the tropopause, at the base of the stratosphere.

All of the atmosphere above the tropopause is heated primarily by solar ultraviolet radiation from above. Each step of the Chapman process adds heat. We will soon see that concentrations of ozone vary rapidly with time and in three-dimensional space. Therefore, the spatial distribution of ozone in the stratosphere shows where local heating is happening, and the concentration of ozone in the stratosphere shows how much local heating is happening provided sufficient solar ultraviolet radiation is available. Measurements of "total column ozone" at any of several hundred stations around the world vary by the minute. These variations show close relationships not only to changing seasons, but also to changing weather.

How Is Ozone Distributed by Latitude and Season?

In 1924, Gordon Dobson built the first instrument to measure total ozone in a column of air extending from Earth's surface into space. By 1958, he had established a network of these instruments throughout the world. Due to his leadership, ozone amount is now measured in **Dobson Units**[119] (DU). Today, data on total column ozone from instruments

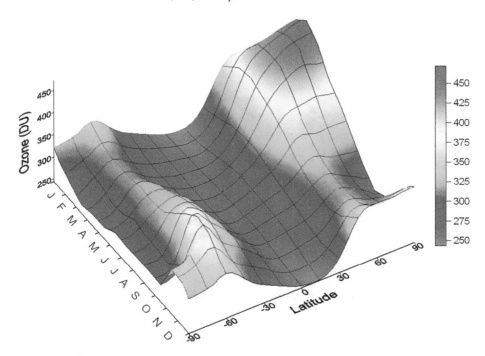

Figure 6.3 The concentration of ozone as a function of month of the year and latitude, showing how ozone accumulates near the poles in late winter and early spring. Monthly amounts are estimated from ground-based data measured between 1964 and 1980.

at more than 500 ground stations around the world are readily available online,[120] as are extensive data from satellites, weather balloons, etc.

The average monthly amounts of ozone, in Dobson Units, measured from the ground between 1964 and 1980 are shown in Figure 6.3 as a function of latitude and month of the year.[121] Note that there is more ozone above the poles than in the tropics during late winter and early spring (February, March, and April in the Northern Hemisphere and August, September, and October in the Southern Hemisphere). During the first half of these time periods, the amount of solar ultraviolet radiation per unit ground area is decreasing due primarily to the decreasing angle of Sun above the horizon. Photodissociation of ozone in the Chapman cycle reaches a low, and therefore ozone concentrations increase. During the latter half of these time periods, the amount of solar ultraviolet radiation per unit ground area is increasing as Sun rises more above the horizon. Photodissociation of ozone increases, and therefore concentrations of ozone decrease.

More ozone absorbs more ultraviolet radiation, warming the ozone layer and cooling Earth. The late-winter, early-spring polar concentrations of ozone help cause the lowest minimum temperatures in polar regions, especially in the Arctic. The primary effect of the increase in ozone depletion since 1970 has been, as described in the next chapter, an increase in minimum temperatures.

Depletion of Ozone by CFC Gases

It is during these late-winter, early-spring peaks in total column ozone that ozone is depleted by **chlorofluorocarbon** gases (CFCs). CFCs were invented in the 1920's, were manufactured especially during World War II, and became very popular in the 1960s for use as refrigerants, spray-can propellants, solvents, and foam blowing agents because they are very inert and therefore do not interact chemically with most other substances. By 1970, a wide variety of substances could be bought in spray cans, such as paints, hair spray, cooking oils, solvents, and lubricants. CFCs were the favored propellants, and their emissions into the atmosphere began to increase substantially by 1970, as shown in Figure 3.3 on page 37.

In 1974, Mario Molina and F. Sherwood Rowland discovered[122] that CFCs are broken down by solar ultraviolet radiation in the upper stratosphere, releasing chlorine atoms. A chlorine atom reacts with an ozone molecule to form a molecule of chlorine oxide and a molecule of oxygen. The molecule of chlorine oxide, then, reacts with an oxygen atom to form a chlorine atom and a molecule of oxygen. Since chlorine atoms are not consumed as a net result of these two chemical reactions, a single chlorine atom can lead to the photodissociation of more than 100,000 molecules of ozone. In this way, chlorine atoms catalyze chemical reactions related to the Chapman cycle. A catalyst is a substance that is not consumed in a sequence of chemical reactions but that causes the reactions to occur faster and with less energy input.

While chlorine atoms can be released from the gas phase of CFCs, the release of chlorine is enhanced dramatically in the presence of **polar stratospheric clouds** (PSCs). These clouds form in the stratosphere at altitudes above Earth of 9 to 16 miles (15 to 25 km) when temperatures in the depth of winter go below –78°C (–108°F or 195K). These clouds are more prevalent above Antarctica, where temperatures are lower and remain lower for longer times, than above the Arctic. Starting in mid-winter, solar ultraviolet radiation increases, driving chemical reactions on the surfaces of the cloud particles that dissociate CFCs, thereby releasing chlorine atoms. Qing-Bin Lu suggests that cosmic rays also play a role.[123]

It takes time for CFCs to rise from where they are used at Earth's surface to the stratosphere. This delay accounts for the lag times between the peak of the atmospheric chlorine curve (green line) and the peaks in ozone depletion (black line) and temperature anomalies (red bars) in Figure 3.3 on page 37. CFCs can remain in the atmosphere for as long as a century.[124]

The Polar Jet Stream and the Polar Vortex

Since there is warming where there is ozone, the distribution of ozone has major effects on, and is affected by, the **polar jet stream** and the **polar vortex**. Both have been in the news in recent years, connected with unusually severe weather in mid-latitudes.

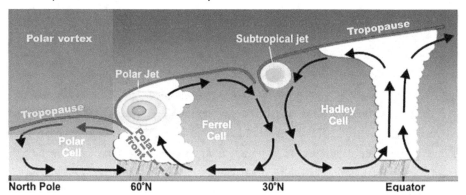

Figure 6.4 The polar jet stream forms just below the tropopause (red line) and above the polar front, where warm, moist air from the mid-latitudes meets cold, dry air from the Arctic. The polar vortex forms during the winter above the polar atmospheric circulation cell.

There is much more detail known about the distribution of ozone in polar regions in late winter to early spring than is shown by the average concentration values plotted in Figure 6.3. Figure 6.4 shows a cross-section of Earth's lower atmosphere from north pole to equator. In the troposphere, warm, moist air from the mid-latitudes moves north to rise over colder, drier, denser air from the Arctic along the polar front (dashed red line).[125] The resulting low pressure near Earth's surface and high pressure at higher elevations form the polar jet stream above the polar front at altitudes of 4.3 to 5.6 mi (7 to 9 km). The jet stream is simply a narrow air current that flows from west to east above the polar front at velocities often approaching 100 km/hr (60 mi/hr) and maximum velocities approaching 400 km/hr (250 mi/hr). The polar jet stream circles Earth near 60°N, but typically wanders both northward and southward, forming so-called "Rossby waves" that have a major effect on storms in the northern hemisphere. A similar polar jet stream circles Antarctica near 60°S, likewise having a major effect on storms in the southern hemisphere.

In winter, the polar vortex forms above the polar cell of atmospheric circulation at elevations from the mid to upper troposphere all the way up to the **mesosphere**, more than 31 mi (50 km) above Earth. This cold-core, high-pressure cyclonic system is bounded by the polar jet stream near the tropopause at 5.6 mi (9 km) altitude and also by the **polar night jet** that forms during winter at altitudes closer to 15 mi (24 km). Warmer air cannot flow into the polar vortex because of circulation patterns around the polar front (Figure 6.4), so that the vortex contains very cold air from the upper atmosphere that typically contains low concentrations of ozone and CFCs. The polar vortex typically spans an area over the pole 600 to 1200 mi (1000 to 2000 km) across and is strongest in winter, causing polar winds from the west to become stronger than the winds that form the polar jet stream. When the vortex is strong, the winds reach the surface, bringing air warmed by the ocean onto land. When the vortex weakens, the polar jet stream buckles, causing significant outbreaks of cold air into lower latitudes. During the summer, the polar vortex weakens considerably, with winds generally confined to the troposphere.

The polar vortex over Antarctica is more pronounced, symmetric, and persistent than the vortex over the Arctic. This is due dominantly to colder temperatures that last for longer periods of time and to the presence of fewer mountains to cause turbulence in the flow. The Antarctic polar vortex has also become stronger, colder, and more persistent since 1979 as ozone depletion increased.[126] The polar vortices are strengthened by increased ozone depletion, which brings decreasing amounts of ozone and consequent decreased heating of the stratosphere. Tropical volcanic eruptions that deplete ozone, as described in Chapter 7, also appear to enhance the Arctic vortex.[127] Countering these effects to some extent, the

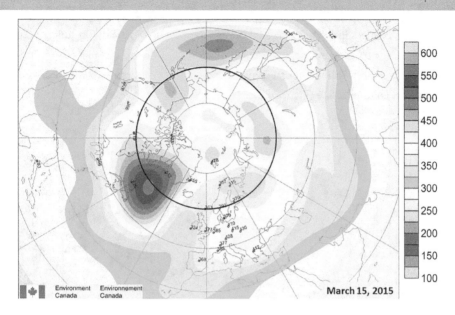

Figure 6.5 A daily map of ozone concentration in the northern hemisphere showing how the most ozone tends to circle the pole in the vicinity of 60°N (black circle), the nominal latitude of the polar jet stream. Colors are ozone concentrations in Dobson Units.

polar vortices bring air from the upper stratosphere, where the concentration of CFCs is lower, down closer to Earth's surface, slowing ozone depletion within the polar vortex.

Winter concentrations of ozone tend to be highest around the edges of the polar vortex, as shown in Figure 6.5[178], and their spatial distribution changes rapidly, as shown by animations and daily ozone maps.[129] If the jet stream encircles a region 600 mi (1000 km) across and moves at 60 mi/hr (100 km/hr), then a molecule of ozone could circle the pole in 31 hours, or more than 6 times within its average lifetime of 8.3 days. Since heat is generated in the ongoing formation and destruction of ozone, high concentrations of ozone at these latitudes warm the jet stream.

The polar vortex, with stratospheric winds from the west in winter, breaks down by summer, when the stratospheric winds blow dominantly from the east. Breakdown of the vortex can occur during the winter but always occurs in the spring, typically in late March or early April in the Arctic, causing sudden stratospheric warming by 54 to 90°F (30 to 50°C). The daily maps of ozone concentration begin to show much less variation with latitude and longitude when Sun returns to polar latitudes.

Sudden changes in the height of the tropopause are also observed. On June 19, 2004, for example, the tropopause over Montreal, Canada, was observed by radar to drop from 8 to 5 mi (13 to 8 km) within 5 hours, contemporaneously with a local increase of more than 20% in total column ozone (Figure 6.6).[130]

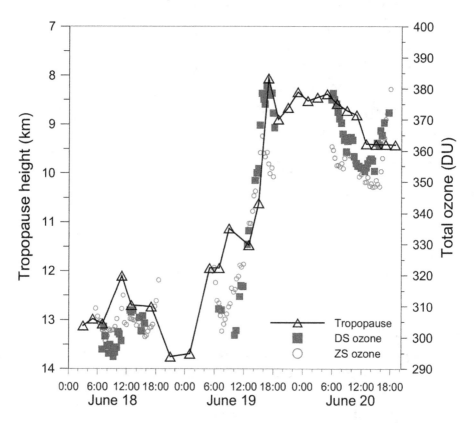

Figure 6.6 On June 19, 2004, tropopause height above Montreal, Canada, was observed to drop 5 km within 5 hours at the same time that total column ozone increased by more than 20%.

Ozone concentrations in the stratosphere are changing constantly, affecting or being affected by changes in atmospheric winds and changes in the tropopause, which separates air heated by Earth from below and air heated by Sun from above.

Ground-Level Ozone

Ozone also occurs in the troposphere, mostly just above the Earth's surface. Ground-level ozone is a colorless, highly reactive, highly irritating oxidant gas that forms when nitrogen oxides (NO$_x$), carbon monoxide (CO), and volatile organic compounds (VOCs) react in the atmosphere with ultraviolet sunlight. Ground-level ozone and particulate matter are the two primary components of "smog" or haze. **Sulfur dioxide** and nitrogen oxides are the two primary components of acid rain. Primary **anthropogenic** (man-made) emitters of NO$_x$ and VOCs are industrial facilities, electric utilities, motor vehicles, gasoline vapors, and chemical solvents, including such seemingly innocuous products as windshield washer fluid.

Ground level ozone forms primarily in highly populated and industrialized areas worldwide, as shown in Figure 6.7 in **parts per billion**.[131]

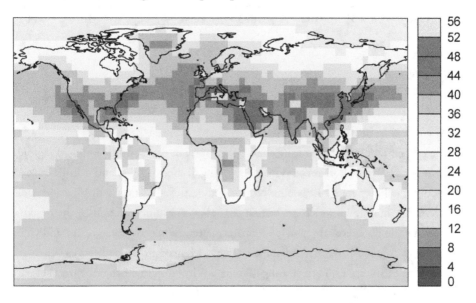

Figure 6.7 Present-day ground level ozone is highest in regions with major populations and industrialization.

Concentrations are highest during hot, sunny, summer days and are especially increased during heat waves, when plants absorb less ozone. Ozone damages plants, causing reduced growth and decreased yield. Efforts to reduce **pollution** have reduced ground-level ozone concentration in the United States by 33% since 1980 (Figure 6.8).[132]

When stratospheric ozone is depleted, more ultraviolet-B radiation reaches the lowermost atmosphere, where it can enhance the ground-level ozone cycles and warm the air, especially in the most industrialized and populated areas. This effect may partially explain why, in Figure 3.2, temperature increased as long as ozone depletion was increasing but stopped increasing soon after ozone depletion began decreasing.

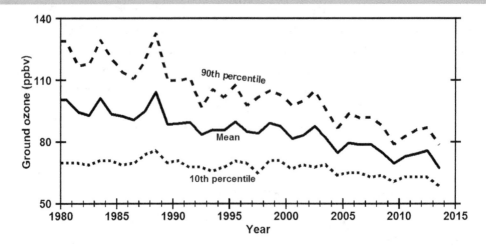

Figure 6.8 Annual range of ground ozone concentrations in parts per billion observed in the United States since 1980 by the Environmental Protection Agency.

How Minute Amounts of Ozone Help Control Weather

Ozone concentrations are constantly changing by time of day, season, latitude, longitude, angle of Sun above the horizon, seasonally changing distance to Sun, and decadal cycles in solar output, all because ozone is both created and destroyed primarily by solar ultraviolet radiation, the intensity of which can be reduced at lower altitudes not only by these factors, but also by clouds, **aerosols**, and particulate matter. While we can average individual observations to get an overview of these changes, we need to remember that when and where ozone exists, atmospheric heating is going on. It is important to recognize that parts per million and even parts per billion of ozone can affect weather.

Around 1920, Gordon Dobson recognized from his study of meteor trails through the atmosphere that temperature increases above the tropopause.[133] He inferred that the cause of this warming must be absorption of ultraviolet radiation by ozone. By 1929, Dobson had shown that "*maximum positive deviations of daily values [of temperatures] from the monthly means are generally found to the rear of surface low-pressure areas, while maximum negative deviations are found to the rear of surface highs. More recently, on the basis of more extensive measurements, Dobson and others refined the earlier results and found that for many occlusions, the maximum positive deviations occur directly over the surface low rather than to the rear.*"[134] Dobson found that the total amount of ozone in the lower stratosphere correlated positively with temperature and potential temperature and negatively with density and the height of the tropopause.

Dobson and others had little evidence for physical or chemical processes that might change ozone concentrations directly other than the dynamic effects of synoptic weather systems scattering and concentrating ozone. With the advent of satellite systems, it is now possible to observe these variations with increasing precision. What is becoming clear is that variations in ozone concentrations may be caused, in part, by dynamic changes in the atmosphere, but these changes, in turn, are partially caused by changes in the concentrations of ozone. There is much detail to work out, but it is clear that ozone plays a major role in weather and in the long-term weather patterns that we call climate.

CHAPTER 7

HOW DOES TEMPERATURE CHANGE WITH OZONE DEPLETION?

"The action of heat is always present, it penetrates all bodies and spaces, it influences the processes of the arts, and occurs in all the phenomena of the universe."
— **Joseph Fourier**, 1822

Theory and observation show that when the concentration of ozone in the **ozone layer** is reduced (depleted), more ultraviolet-B **radiation**, primarily with frequencies between 900 and 1000 THz (Figure 5.4 on page 75), reaches Earth's surface, warming Earth instead of warming the ozone layer. The greatest amount of global warming observed on Earth since 1965 was located below the **Antarctic ozone hole**, where the greatest amount of ozone depletion was also observed at the same time. The second greatest amount of global warming observed since 1965 was in the Arctic, the site of the second greatest amount of simultaneous ozone depletion. The amount of ozone depletion and the amount of global warming are also observed to be closely related, although less severely so, at mid-latitudes.

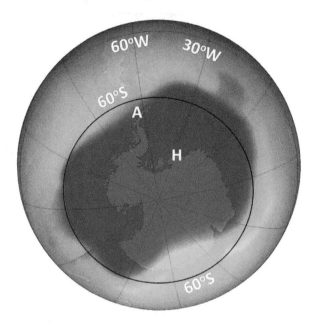

Figure 7.1 The Antarctic ozone hole, observed by satellite on September 24, 2006, covers almost all land and ocean south of 60 °S. Green represents ozone levels around 300 DU; purple around 150 DU, representing 50% ozone depletion.

The Antarctic Ozone Hole

Total column ozone has been measured from the ground, at the British Antarctic Survey stations in the Argentine Islands (A in Figure 7.1)[135] and at Halley Bay (H) since 1957. From 1957 to 1973, ozone values in March (early fall) and October (spring) showed little difference. By the early 1980s, however, spring values were much lower than values measured in the early fall.[136] These observations provided the first unambiguous evidence of the greatest ozone depletion ever observed anywhere on Earth in the form of what became known as the "Antarctic ozone hole." In 1979, the maximum areal extent of the hole was 1.1 million km² (0.4 million mi²). On September 24, 2006, the ozone hole reached its largest areal extent ever observed, covering 30 million km² (11.6 million mi²), or almost all of the land and ocean south of 60 °S (Figure 7.1). By 2014, the areal extent had decreased 20% to 21 million km² (8.1 million mi²). Meanwhile, the minimum total ozone column in 1979 was 194 DU; in 2006, it was 84 DU; and by 2014, it had increased 36% to 114 DU.[137] The area of the Antarctic ozone hole and the **polar vortex** that surrounds it continue to decrease slowly in size as the concentration of ozone in the atmosphere slowly continues to increase.

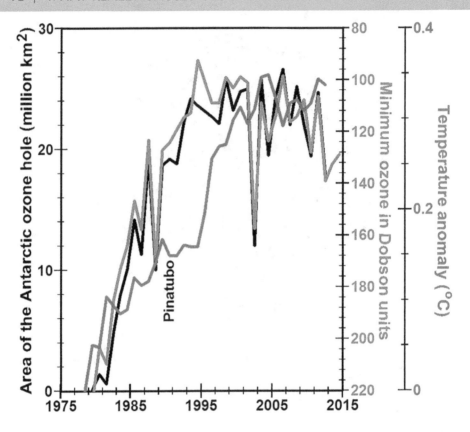

Figure 7.2 The area of the Antarctic ozone hole (black) grew rapidly from the 1970s to the 1990s while minimum ozone (green) in the hole decreased. Mean annual temperature anomaly in the southern hemisphere (red line) is smoothed with a centered 5 point running average. Note powerful cooling effect of Pinatubo eruption.

Since 1979, the area of the Antarctic ozone hole has increased[138] (black line in Figure 7.2), while the minimum amount of ozone each year decreased[3] (green line; note inverted y axis) and the temperature anomaly increased (red line).[139] The divergence of the **temperature** line from 1992 to 1995 was caused by global cooling following the eruption of Mt. Pinatubo in the Philippines in 1991, which will be discussed in Chapter 8.

The Antarctic ozone hole formed in the ozone layer, dominantly at altitudes between 8 and 14.2 mi (13 and 23 km), where ozone **partial pressure** dropped from 15.6 mPa (megapascals) in August (late winter) to essentially zero in October (spring) (Figure 7.3).[140]

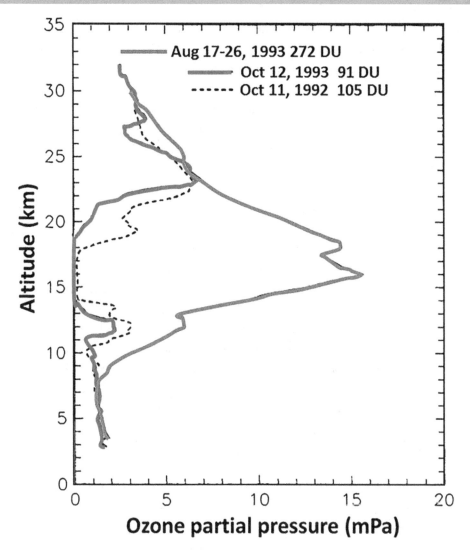

Figure 7.3 Partial pressure of ozone above the South Pole measured from balloons shows that the substantial presence of ozone in August, 1993, disappeared at altitudes from 14 to 19 km (8.7 to 11.8 mi) in October.

Depletion of ozone caused by **anthropogenic chlorofluorocarbons** (CFCs) is thought to occur primarily in **polar stratospheric clouds** that form when temperatures drop below -108.4°F (-78°C) within the polar vortex. The vortex, centered over Antarctica, forms in May at latitudes greater than 60°S and dissipates in October. The Antarctic ozone hole typically reaches its maximum extent during September and its lowest values of ozone in late September to early October. By 1989, ozone levels each October at Faraday/Vernadsky Research Base in the Argentine Islands near the Antarctic Peninsula (A in Figure 7.1) had become depleted by 45% compared to average levels from 1957 through 1970 (350 DU).[141] Minimum yearly ozone levels throughout Antarctica have typically been depleted 45 to 55% relative to levels of 225 DU in 1979 when satellite measurements began. At the same time, the vortex has become "stronger, colder, and more persistent".[142]

The largest warming trend in the world observed between 1976 and 2000 was along the Antarctic Peninsula. Minimum monthly temperatures at Faraday/Vernadsky Research Base increased by 12°F (6.7°C) from 1951 to 2003,[143] representing the greatest warming of this region, according to ice-core studies,[144] in more than 1800 years. These rapid increases in temperature were strongly correlated with decreases in total column ozone. During summer months when ozone is not depleted, maximum monthly temperatures have changed very little since observations began. Between 1958 and 2010, annual mean temperatures increased 5.4°F (3°C) at Faraday/Vernadsky station and 4.3°F (2.4°C) at Byrd Station (80°S, 199.5°W) compared to 1.3°F (0.7°C) globally.[145]

Decreased ozone allows more ultraviolet solar radiation to reach Earth's surface, where it is absorbed most efficiently by ice-free water, a band of which hundreds of kilometers wide exists off the ice-bound coast of Antarctica but still within the Antarctic ozone hole.[146] Summer surface temperatures of the Bellingshausen Sea rose 1°C,[147] the deep water of the Antarctic circumpolar current warmed,[148] and formation of cold Antarctic bottom water decreased substantially.[149] "Southern oceans have warmed at roughly twice the rate of global mean ocean."[150]

Along the Antarctic Peninsula, winter sea ice decreased 10% per decade and shortened in seasonal duration.[151] More than 85% of the marine glaciers in this region retreated, many collapsing into the ocean following the loss of seven very large ice shelves.[152] Warming of interior Antarctica was slowed by the high mean **albedo** (reflectivity) of Antarctic snow (~0.86), nearly twice the albedo of Arctic snow, by the decrease in solar flux approaching the South Pole, and by the polar vortex.

Arctic Amplification

The second greatest increase in surface temperatures was in the Arctic.[153] Ozone is depleted primarily during December, January, and February within the Arctic polar vortex, but depletion spreads south to mid-latitudes in spring, enhanced by increasing nitrogen-oxide-related "summer" depletion.[154] The asymmetric distribution of mountains, land, and ocean throughout the Arctic makes this vortex much more variable than the Antarctic vortex on daily to inter-annual time scales. The greatest amounts of total column ozone occur near and just outside the edge of the polar vortex in the **polar night jet**, located throughout the **stratosphere**. Ozone concentrations are continually changing, as shown in the daily ozone maps.[155] Ozone depletion during winter/spring has been increasing in the Arctic since the

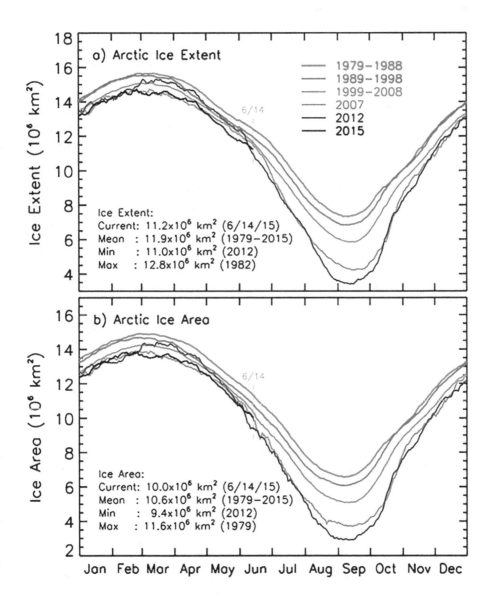

Figure 7.4 The extent (a) and area (b) of Arctic Sea Ice have been decreasing rapidly since 1979.

1950s, exceeding 80% at altitudes of 11 to 12 mi (18 to 20 km) in early 2011, comparable for the first time with the Antarctic ozone hole.[156]

Ground-based temperatures north of 65°N increased at a rate of approximately twice the global average from 1965 to 2005.[157] Between 1966 and 2003,[158] annual mean land-surface temperatures north of 60°N increased 1.5°C compared to 0.6°C for the northern hemisphere as a whole. All this warming is unprecedented within the past 600 years.[159] Satellite data from 1981 to 2005 for all areas north of 60°N show an average increase of 3.1°F (1.7°C) with increases of 5.2°F (2.9°C) over Greenland, 3.6°F (2.0°C) over North America, 2.3°F (1.3°C) over sea ice, and 0.54°F (0.3°C) over Europe.[160] The average increase in monthly mean temperature from 1966 thru 2010 was more than 9°F (5°C) north of 70°N, more than 7.2°F (4°C) from 65° to 70°N, more than 2.7°F (1.5°C) from 55° to 60°N, and for the months of June through September was more than 3.6°F (2°C) for all stations on land throughout the northern hemisphere.

The extent of Arctic sea ice,[161] shown in Figure 7.4, declined at rates of more than 11% per decade since 1979[162] and reached a record low on September 16, 2012, nearly 50% lower than the average extent between 1979 and 2000. The extent of terrestrial snow cover in June has decreased 17.8% per decade since 1979.[163] Average snow-covered area in the northern hemisphere decreased approximately 7%, primarily since 1982. Loss of ice in Greenland has been accelerating at a rate of 21.9 gigatons per year squared.[164] The

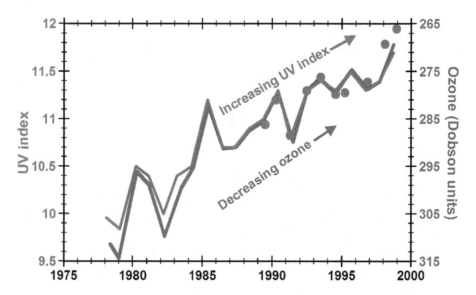

Figure 7.5 The calculated UV index (red) increases as the observed total column ozone concentration decreases.

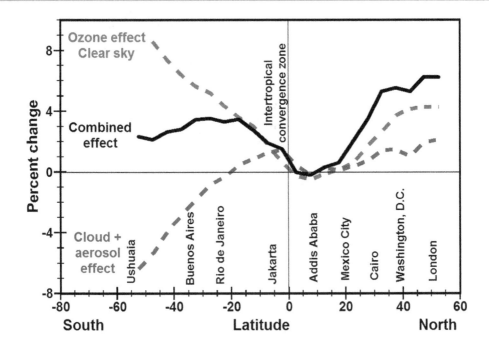

Figure 7.6 Sun-burning radiation (black line) increased between 1979 and 2008 at almost all latitudes, with the greatest increases in the northern hemisphere.

Canadian Arctic Archipelago has been losing ice at a rate of 61 gigatons per year.[165] Ice-cap melt rates on Ellesmere Island in the last 25 years have been the highest observed in 4200 years.[166]

Mid-Latitudes

When ozone is depleted, a narrow band of ultraviolet energy with frequencies between 900 and 1000 THz—as shown in Figure 5.4 on page 75—reaches Earth's surface. This band, known as erythemal radiation or sunburning radiation, has been measured and studied in considerable detail from a health perspective.[167] For every 1% decrease in ozone in the stratosphere, an additional 2% of sunburning radiation reaches Earth.[168] The sensitivity of biologic organisms to **ultraviolet radiation** increases rapidly as the frequencies increase from 937 to 1000.[169]

The strength of ultraviolet (UV) radiation, the UV index, can be measured at a specific place and time and can be calculated based on total column ozone in the atmosphere above that point. Ozone measured during summers from 1978 to 1999 over Lauder, New

Zealand (45°S)[170] decreased, as shown by the blue line (y-axis reversed) in Figure 7.5, while the predicted UV index (red line) increased. The red dots show actual observed UV indices.

The percent change in sunburning radiation from 1979 to 2008 is shown as a function of latitude[171] by the black line in Figure 7.6.

The dashed red line shows the effect of ozone depletion for a clear sky. The dashed blue line shows the effect of clouds and **aerosols**. The combined effect (black line) is greatest in the northern hemisphere, agreeing with the observation that global warming, during the same period of time, was greater in the northern hemisphere than in the southern.

Global changes in climate since 1965 are closely linked to observed global changes in emissions of CFCs that deplete the ozone layer, as explained so far in this book. Volcanic eruptions of all sizes also deplete ozone and have played a major role in determining regional changes in climate during the same time period. We will explore the details of the volcanic connection in the next chapter.

CHAPTER 8

HOW CAN VOLCANOES BOTH COOL AND WARM EARTH?

"In science, self-satisfaction is death. Personal self-satisfaction is the death of the scientist. Collective self-satisfaction is the death of the research. It is restlessness, anxiety, dissatisfaction, agony of mind that nourish science."
—**Jaques Monad**, 1976

n late 2006, while working on something unrelated to climate, I discovered data from the layers of ice beneath Summit Camp, Greenland, showing that since the coldest part of the last ice age, around 24,000 years ago, ice layers containing the greatest amounts of volcanic **sulfate** per century[172] also contained **oxygen isotope** evidence for the greatest rates of global warming per century.[173] Sulfate is created from **sulfur dioxide**, which is erupted in large quantities by **volcanoes**. These data appeared to show that the largest amounts of volcanism occurred precisely during the two periods of most rapid increases in temperature: the **Bølling warming** from 14,600 to 13,000 years ago, when sea-level rose more than 66 ft (20 m) within a few hundred years, and the **Pre-boreal warming** from 11,950 to 9,375 years ago, when the world finally warmed out of the last ice age. I wondered—could volcanic eruptions have caused the warming?

That conclusion seemed inescapable, but volcanologists and climatologists know that nearly all major **explosive** volcanic eruptions described in written history have been followed by several years of global cooling of approximately 1.0°F (0.6°C). This cooling is readily explained by observations, made since the mid-20th century, of large eruptions typically ejecting 5 to 20 megatons of sulfur dioxide into the lower **stratosphere**, where it spreads around the globe within months as it is slowly oxidized and hydrated to form **sulfuric acid**. Sulfur dioxide, because of its low vapor pressure, quickly attaches to airborne dust particles, forming, over time, an **aerosol** with particle sizes that grow large enough to reflect and scatter sunlight, thereby cooling Earth.

How, then, could volcanic eruptions cause both global cooling and the greatest rates of global warming observed in the past 130,000 years? I spent the better part of a year examining in layer-by-layer detail these high-quality data, which were collected meticulously by an outstanding team of scientists (see Figure 8.1). Because sulfate can be deposited not only from volcanic eruptions, but also from sea salt and continental dust, the researchers found a way to subtract these typically minor non-volcanic contributions from total measured sulfate based on their associated concentrations of sodium and calcium.[174] This method is objective, based on sound chemical theory, and therefore provides a statistically accurate estimate of the remaining volcanic sulfate. More than 63% of the 7000 measured ice layers formed in the past 25,000 years contained zero volcanic sulfate. The ratio of data signal to background noise is high: peak sulfate per century (2028 **parts per billion** [ppb]) during the maximum warming event since the **last glacial maximum**, between 25,000 and 21,000 years ago, is 218 times greater than average sulfate per century (9.3 ppb) and 441 times greater than average sulfate (4.6 ppb) between 5000 and 1000 years before the present.

Changes in oxygen isotopes ($\delta^{18}O$, pronounced "delta-oh-eighteen") are measured in shells of miniscule **foraminifera** and in air bubbles trapped in ice. They are widely accepted as proxies (substitutes) for mean **paleotropical** sea-surface **temperature** and global ice volume, and hence for paleotemperature in Greenland ice. The $\delta^{18}O$ data shown in red in Figure 8.1 include 16,600 samples collected from air bubbles in sulfate-rich ice layers, ensuring both high accuracy and high resolution in temporal correlation.

There is a close, layer by layer correspondence between increasing temperature and increasing sulfate, especially when high levels of sulfate occur in many contiguous layers. Within the time interval represented in Figure 8.1, each sampled ice layer represents a range of two years of snow accumulation, on average, since 10,000 years ago, to more than

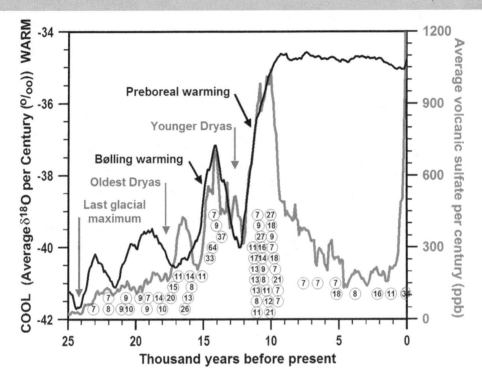

Figure 8.1 Layers of ice beneath Summit, Greenland, containing the largest concentrations of volcanic sulfate per century (red) also contain oxygen isotopic evidence for the greatest rapid melting of continental glaciers (black) at the close of the last ice age. The numbers in blue circles are the number of contiguous ice layers sampled that contained volcanic sulfate and thus represent periods of relatively continuous volcanism. Periods of maximum cooling back into ice-age conditions are shown by blue arrows; periods of major warming by black arrows.

eight years per layer before 14,000 years ago. Sulfate is consistently high in most ice layers formed from 11,950 to 9,375 years ago, implying relatively continuous volcanism as the ocean warmed out of the last ice age.

I began to dig deeper. I had worked throughout Iceland in 1967 and 1968, preparing my PhD thesis. I knew that Iceland is dotted with basaltic **table mountains**, also called **tuyas** (Figure 8.2).[175]

Basalt that erupts under a glacier cannot flow downhill because it is cooled rapidly by the ice. Therefore, the fresh basalt builds vertically until it reaches the surface of the glacier (Figure 8.3).[176]

Figure 8.2 Herðubreið, a tuya or table mountain, in northeastern Iceland.

Careful dating of these rocks shows that "12 of the 13 dated table mountains experienced their final eruptive phase during the last glaciation." Clearly, basaltic volcanism was prevalent in Iceland precisely when the world was finally warmed out of the last ice age.

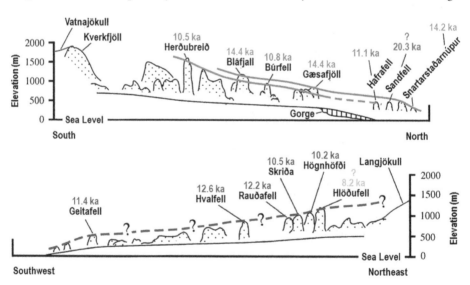

Figure 8.3 Ice sheet surface elevations in Iceland at different times, in thousand years before present (ka), based on concentrations of cosmic-ray-produced helium 3 isotopes on the rocks erupted at the tops of tuyas along profiles in northern and southwestern Iceland.

Confronted by all of these apparently reliable observations, my scientific curiosity was at a fever pitch. There was something going on here concerning global warming that we did not understand. I recognized that resolving this enigma might provide an important contribution to one of the largest scientific and political issues of our time. Since 2006, when I began working on this problem, I have stepped back many times to try to assess objectively whether I was heading in the right direction. Each time I revisited the sulfate data from the Greenland ice, however, I concluded that these data were just too clear and reliable to be ignored. In science, good data are as close as we can get to ground truth.

What I have concluded is that all volcanic eruptions deplete the **ozone layer**, causing global warming, but that major explosive eruptions also cause aerosols to form in the lower stratosphere that reflect and scatter sunlight. The cooling from the aerosols is much stronger than the warming from ozone depletion. Hence, explosive volcanoes that form aerosols cool Earth, while **effusive** and quietly degassing volcanoes, that do not form aerosols, warm Earth. Let's look at the details.

How Do Explosive Volcanic Eruptions Cool Earth?

On June 15, 1991, Pinatubo volcano in the Philippines explosively erupted (Figure 8.4)[177] approximately 5 km³ of viscous dacitic **magma**, most of it within 9 hours, projecting 17 Mt (megatons) of sulfur dioxide as much as 22 mi (35 km) into the stratosphere, with peak mixing ratios observed via the SAGE II satellite of 300 ppbm (parts per billion by mass), decreasing to 160 ppbm within 180 days, and remaining above background concentrations for two years.[178] Within 21 days, sulfur dioxide injected into the lower stratosphere had circled Earth and spread poleward to 30°N and 10°S. Within one year, the sulfur dioxide had spread throughout the globe. Approximately 13 Mt of erupted sulfur dioxide ultimately were oxidized and hydrated to form an aerosol, made up of 75% sulfuric acid and 25% water, primarily at altitudes[179] between 12 and 16 mi (20 and 25 km). It took as long as 13 months for the aerosols to form in the Arctic.[180]

Optical depth is a normalized logarithmic measure (i.e., 0.0 is the standard) of how much sunlight is absorbed in the atmosphere. The optical depth of the atmosphere for mid-visible (green) light at 600 THz increased to 0.3 within 2 months, peaked at 0.4 by late 1992, and averaged globally 0.1 to 0.15 for 2 years. This excursion in optical depth is thought to have decreased solar **radiation** at Earth's surface by approximately 2.7 W/m² (watts per square meter) in August 1991, and by about 2.5 W/m² by the end of the year.[181] Surface temperature decreased up to 1°F (0.6°C) below normal in the northern hemisphere and averaged 0.7°F (0.4°C) below normal over most of the globe through 1993.[182]

During the winters of 1991 and 1992, however, surface temperatures increased by as much as 5.4°F (3°C) above normal over North America, Europe, and Siberia (Figure 8.5) due to ozone depletion.[183] The lower stratosphere warmed by as much as 5.4°F (3°C) by mid-November 1991 but cooled by 0.6°F (0.35°C) with respect to pre-eruptive levels by early 1992.[184]

Most explosive volcanoes erupt magmas that are more evolved, meaning that these magmas have had enough time residing in Earth's upper layers to assimilate crustal components that alter them from basalt—the most primitive magma—to rock types such as andesite, **dacite** (a quartz-rich andesite), and **rhyolite**. Evolved magmas are more viscous than primitive basalt and have lower temperatures, features that cause them to erupt explosively. When they erupt, they eject megatons of sulfur dioxide within hours into the stratosphere where, over weeks to months, the sulfur dioxide gas combines with oxygen and water vapor to form a haze of sulfur dioxide that coalesces onto condensation nuclei, forming aerosols with particle sizes dominantly in the range of 300 to 500 nanometers.[185]

Figure 8.4 On June 15, 1991, the explosive volcano Pinatubo in the Philippines erupted 1.2 mi³ (5 km³) of dacitic/andesitic magma as much as 22 mi (35 km) up into the atmosphere. The main eruption lasted 9 hours.

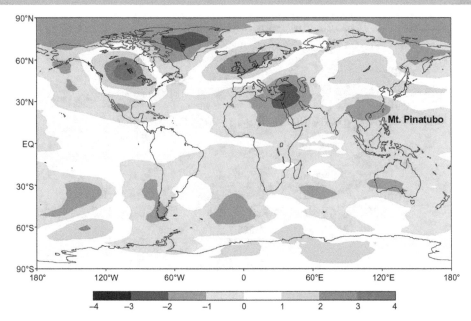

Figure 8.5 Surface temperature anomalies during the winter following the eruption of Mt. Pinatubo (December 1991 through February 1992) show warming over northern continents just outside of the polar vortex, most likely caused by ozone depletion dominating before the aerosols were fully formed, causing net cooling.

The larger the size of the particle, the lower the frequency of the radiation with which the particle can interact. When they first form, the particles reflect and scatter high-frequency ultraviolet solar radiation—the same radiation that reaches Earth when ozone is depleted—and as they grow in size, they reflect lower-frequency visible light.

This entire process causes global cooling of about 1.0°F (0.6°C) for several years. Formation of these aerosols takes time and depends on the **photochemistry** and stratified nature of the lower stratosphere. In the more turbulent **troposphere**, the process does not occur. The reflection and scattering of solar radiation by stratospheric sulfur dioxide aerosols dominates over the effects of ozone depletion, so that the net effect of explosive volcanic eruptions is to cause global cooling.

How Does a Sequence of Large, Explosive, Volcanic Eruptions Increment the World Into an Ice Age?

The global ocean can retain the effects of this net cooling for more than 100 years, as shown in Figure 8.6, based on computer modeling of ocean temperature.[186] Thus, a sequence of explosive eruptions can have a net cooling effect, as suggested in Figure 8.7, a plot of

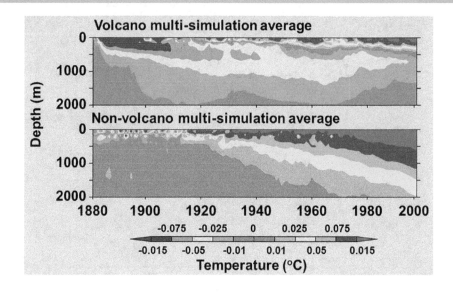

Figure 8.6 Modeled temperature anomalies for a gradually warming upper ocean with (upper plot) and without (lower plot) a cooling effect from the 1883 eruption of Krakatau. Note that the cooling effect appears to persist for more than one hundred years.

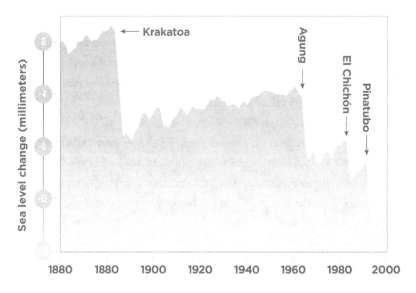

Figure 8.7 Modeled sea level change suggests an accumulation of cooling caused by large volcanic eruptions. Shown here is an annual time series of global-mean sea level change due to thermal expansion.

theoretical sea-level changes caused by changes in ocean temperature, based on computer modeling.[187] Currently, large explosive volcanic eruptions are occurring approximately twice per century. If they were to occur every decade or so for hundreds to thousands of years, they could cool the world incrementally into another ice age.

The origin of ice ages is still a highly debated topic in Earth science. The model most widely accepted as a possible cause is based on **Milanković cycles**, named after the Serbian geophysicist Milutin Milanković, who theorized that variations in the **eccentricity**, **axial tilt**, and **precession** of Earth's orbit change the amount of solar radiation reaching Earth over 100,000-, 41,000-, and 23,000-year cycles. While these Milanković cycles may have some small effect on decreasing global surface temperatures, a sequence of explosive volcanic

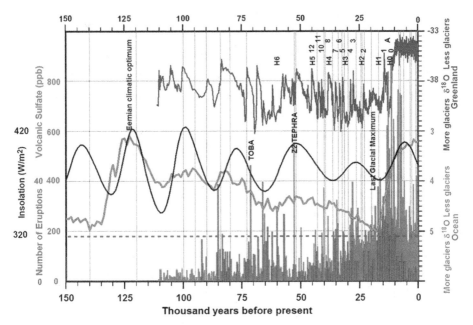

Figure 8.8 Changes in global temperature over the past 125,000 years (green line) and more detailed changes in temperature over Greenland (purple line) show much more frequent change than can be explained by the Milanković cycles (black line). Red bars show the amount of volcanic sulfate measured in the GISP2 borehole under Summit, Greenland. The largest sulfate anomalies clearly coincide with the rapid warming since the Last Glacial Maximum. Blue bars are the number of eruptions per century inferred by Zielinski (1996) to correct for ice compaction. Numbers are Dansgaard-Oeschger sudden warmings. H numbers are Heinrich events of more icebergs in the North Atlantic.

eruptions provides a much more direct explanation that can also explain why changes in temperature are usually rather abrupt and not along smooth cycles.

David Laing[188] has suggested that Milanković cycles might have a gravitational influence on Earth's delicately balanced plate tectonic system. Increased plate motion induces increased volcanism, which would therefore bear the imprint of the Milanković cycles. He notes that in Figure 8.8 there is a clear correspondence between minima in the Milanković cycles (black line) and increased number of eruptions (blue bars).

Figure 8.8 also compares the changes in volcanism and temperature since the **Eemian climatic optimum**, the last interglacial period around 125,000 years ago. There are many different data sets plotted in this figure, and you may find it confusing at first. The purpose of the figure is to allow you to determine for yourself whether you think the observed changes in volcanic activity might provide a more compelling explanation for observed changes in temperature than simply variations in insolation due to Milanković cycles.

The green line, representing temperature, with lowest temperatures down and warmest temperatures up, is the most important line in the figure. This line is based on the oxygen isotope ($\delta^{18}O$) **proxy** for tropical ocean-bottom temperature and is also affected by ice volume.[189] This green line was determined from 38,000 individual $\delta^{18}O$ measurements from ocean sediment cores at 57 sites distributed around the world. The temperature range is approximately 5°F (2.8°C) between the last glacial maxima at 140,000 and 24,000 years ago and the present.[190] Note how the world cooled slowly and incrementally, with brief periods of warming, from 125,000 years ago to the depth of the last ice age, the last glacial maximum, around 24,000 years ago. Then the world warmed suddenly in several stages, exiting the ice age by 10,000 years ago.

The purple line is the $\delta^{18}O$ proxy for temperature and glaciation from the **GISP2** drill hole in central Greenland,[191] showing a record of frequent change that is much more detailed.

The black curve shows the changes in temperature theoretically resulting from variations in solar radiation received by Earth due to Earth's orbital changes, as predicted by Milanković.[192] Note how much more frequently change was actually occurring than could be explained by Milanković cycles alone. Laing further notes that the rough correspondence between peaks and troughs in the Milanković and temperature proxy curves is a good example of the ambiguity inherent in selective correlations. Although it has been widely argued that the correspondence reflects a direct cause-and-effect relationship, it is also possible, as he argues, that the relationship is indirect—i.e., mediated by volcanism.

Red bars show individual sulfate anomalies measured in the GISP2 ice core.[193] The largest sulfate anomalies clearly coincide with the rapid warming since the last glacial maximum around 20,000 years ago. Blue bars show the number of eruptions per century inferred by Zielinski after correcting for ice compaction.[194] The dashed blue line, at 18 eruptions per century, is the approximate boundary between the cooling effects of a small number of explosive eruptions and the warming effects of a large number of effusive eruptions.

The vertical grey lines, which stretch from the bottom to the top of the graph, are known major volcanic eruptions.[195] **Toba** is the largest known explosive volcanic eruption. The Z2 **tephra** is a volcanic **ash** widely recognized in deep-sea sediment cores.

H0 thru H6 are **Heinrich events**, in which large armadas of icebergs broke off from glaciers and floated out into the North Atlantic.[196] Numbers 0 thru 12 and the letter A are **Dansgaard-Oeschger** sudden warming events, discussed further below, in which the world abruptly warmed out of the ice age within years to a decade or two. The warming stopped, however, before the deep ocean could be warmed, so that the still chilly ocean cooled the world back into the ice age within decades to centuries.

There is considerable evidence from the geologic record that major changes in climate are caused by changes in the rates of volcanism that are in turn caused by changes in the motions of the eight major and many minor **tectonic plates** that make

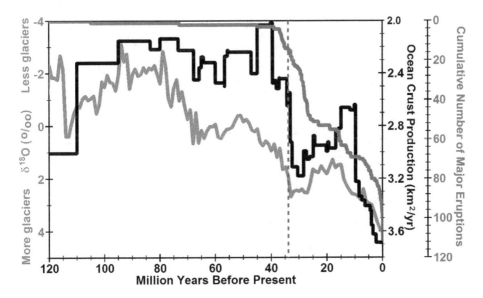

Figure 8.9 The world cooled into an ice age (green line) as the cumulative number of major volcanoes increased (red line, inverted scale) and the ocean crust production (black line, inverted scale) increased, especially at around 34, 12, and 3 million years ago.

up the surface of Earth. The cumulative number of known major explosive eruptions[197] during the past 120 million years is shown by the red line in Figure 8.9, with numbers increasing downward in order better to visualize that increased volcanism is correlated with decreasing temperature. While it gets more difficult to map and date older eruptions—and clearly there must have been many more than the three known eruptions shown in this plot before 40 million years ago (commonly abbreviated in Earth Science as Ma meaning megaanna)—there is clearly a major increase in the rate of known eruptions between 39 and 34 Ma, especially in the Davis Mountains of Trans-Pecos Texas[198] and the upper Rio Grande rift valley of central New Mexico and southern Colorado. This is the same general time period in which the direction of motion of the Pacific Plate appears to have changed from moving northward, as is assumed by most scientists to be shown by the linear trend of the Emperor **Seamount**

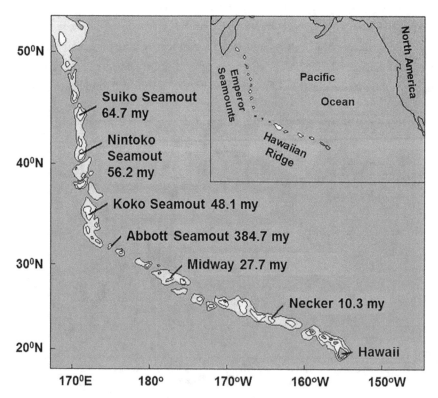

Figure 8.10 *The Emperor seamount chain (north of 32°N) was formed between 85 and 39 million years ago, evidence that the Pacific Plate was moving northward. By 38.7 million years ago the Hawaiian Islands and seamounts (south of 32°N) were being formed, evidence that the Pacific Plate was by then moving more northwestward.*

Chain (Figure 8.10), to moving northwestward, as shown by the linear trend of the Hawaiian chain of volcanoes. It is reasonable to hypothesize that during this time, the rate of **subduction** of the Pacific Plate under Asia increased substantially, but the details are still being discussed.[199]

The black line in Figure 8.9 shows the area of ocean crust formed per year (numbers increase downward),[200] which reached a minimum between 45 and 39 Ma, when the plates were relatively stationary. The sudden increase in ocean crust production around 39 Ma is again consistent with the hypothesis of the onset of major subduction, shoving the Pacific ocean plate under Asia. Since the surface area of Earth remains relatively constant over time, increased ocean crust production at oceanic **spreading rifts** is closely related to increased subduction of ocean plates under continental margins. Increased subduction is associated with increased explosive volcanism in **island arcs**.

The green line in Figure 8.9 is based on oxygen isotopes ($\delta^{18}O$).[201] The range in temperatures shown is approximately 13°F (7°C).[202] The dashed blue line at 33.9 Ma marks a time of major cooling of Earth, leading to sudden, extensive glaciation in Antarctica and major changes in flora and fauna in North America and Eurasia.[203] Note the substantial cooling (green line) during times of major seafloor production (black line) and/or explosive volcanism (red line) at 34, 12, and 3 million years before the present.

It seems clear, from Figure 8.9, that increased rates of explosive volcanism over centuries to millennia cool Earth into ice ages. There is much more evidence throughout the geologic record of increased rates of explosive volcanoes causing rapid cooling, although cooling into ice ages only occurred four times in the last 580 million years (approximately 450 to 370, 350 to 280, 200 to 110, and 40 to 0.01 million years before the present; see Figure 10.6 on page 148).

How Do Effusive Volcanic Eruptions Warm Earth?

Effusive eruptions of primitive basaltic magma, on the other hand,

1. extrude extensive lava flows for days, months, centuries, and even hundreds of thousands of years,
2. emit 10 to 100 times more volatiles, including water, carbon dioxide, sulfur dioxide, and hydrogen chloride, per cubic kilometer of magma than explosive volcanoes,[204]
3. do not eject most of these emissions high enough to reach the stratosphere,
4. do not form extensive stratospheric aerosols, and
5. therefore cause net warming.

The two largest effusive eruptions in recorded history were in South Iceland near the intersection of a developing east-west **transform fault** and the main spreading rift of the mid-Atlantic Ridge.[205] Eldgjá erupted 18 km³ of basalt from 934 to 940,[206] and Laki erupted 15.1 km³ from June 8, 1783, to February 7, 1784.[207] Laki ejected 24 Mt of sulfur dioxide into the lower stratosphere and an additional 96 Mt of sulfur dioxide into the troposphere, where the jet stream carried much of it to the southeast, toward Europe.[208] Severe acid damage to vegetation from Iceland to Eastern Europe and Italy suggests concentrations of sulfur dioxide could have been as high as 1000 ppb, at least three orders of magnitude larger than background concentrations.

Figure 8.11 The Holuhraun lava field on Bárðarbunga volcano in central Iceland, erupted from August 29, 2014, to February 28, 2015, producing more than 85 km² (33 mi²) and 1.4 km³ (0.34 mi³) of basaltic lava—the largest in Iceland since 1783. © *Arctic-Images/Corbis*

A "**dry fog**" from Laki blanketed much of Europe, primarily from mid-June through August 1783. Sulfur dioxide is invisible, but when absorbing **ultraviolet radiation**, the **electronic transitions** cause **fluorescence** in the visible spectrum, explaining the appearance of "dry fog." There were ten major eruptive phases from June thru October. Sulfur dioxide absorbs solar radiation in the visible light range. Because of this absorption by ground-level sulfur dioxide, surface temperatures in Western Europe in July increased by as much as 3.3°C above the 30-year mean centered on 1783, and temperatures in Central

England were the highest recorded over the interval from the first measurements in 1659 until 1983.[209] The lesser amount of sulfur dioxide injected into the stratosphere would not have had adequate time to form cooling aerosols during the many eruptive phases in July.

Unusually high concentrations of ground-level sulfur dioxide were thus the most likely cause of the increased temperatures shown in Figure 8.12.[210] Based on this, I

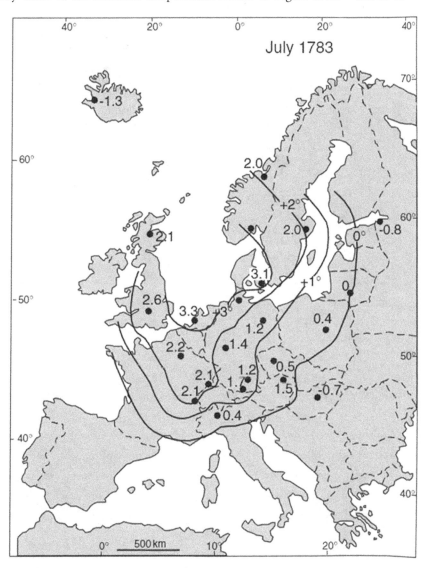

Figure 8.12 Temperatures in Europe increased up to 5.4°F (3°C)—compared to the mean from 1768 through 1798—during July 1783 when a "dry fog" was caused by the ongoing eruption of Laki volcano in Iceland. (Values in °C)

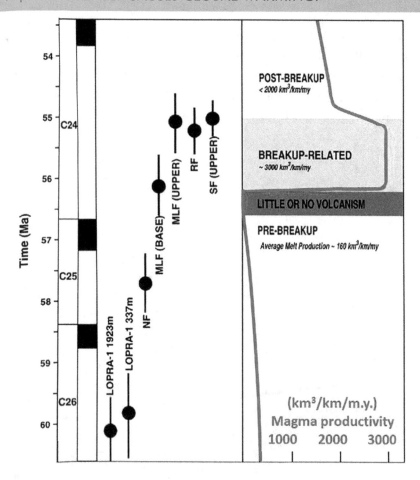

Figure 8.13 A sudden spike in magma productivity (red line) was contemporaneous with the Paleocene-Eocene Thermal Maximum around 56.1 million years ago (Ma).

suggested that longer-term and larger-scale climate warming might also be caused[211] by ground-level sulfur dioxide, but concentrations did not appear to be either high enough or extensive enough. Ultimately, I realized that the concentrations of sulfate and its precursor sulfur dioxide per ice layer or per century measured in ice cores simply indicate the rate of volcanism and the likely amount of associated ozone depletion from erupted chlorine.

As was noted previously, basaltic volcanism under ice forms long, flat-topped, steep-sided **table mountains**, or **tuyas**, found throughout Iceland (Figure 8.2). As was also noted, "12 of the 13 dated table mountains experienced their final eruptive phase during the last deglaciation."[212] The vast magnitude of the eruptions could well have resulted which would have increased magma production by on the top of the magma sources, thus inducing ean **magma chamber**.[213] The relatively continuous, m from 11,950 to 9,375 years before the present, r radiation or more credibly by gravitational stress ankovíc cycles, was apparently sufficient to warm the ocean, and thus Earth, out of the last ice age.

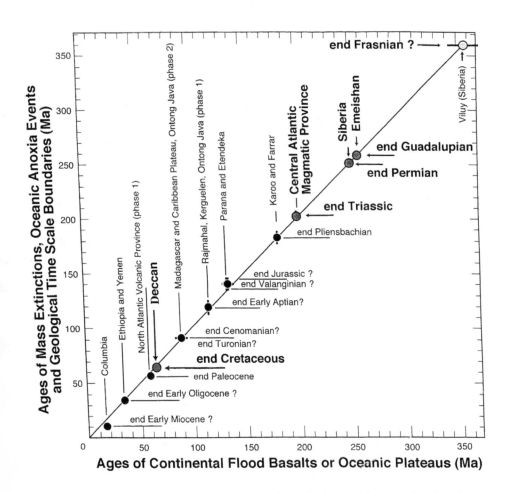

Figure 8.14 Flood basalts labeled vertically above the diagonal line and dated on the x-axis in megaanna (Ma), meaning millions of years before present, typically occur at the same time as mass extinctions dated on the y-axis and labeled horizontally below the diagonal line.

from the melting of overlying ice, which would have increased magma production by reducing the vertical load pressure on the top of the magma sources, thus inducing further melting within the subterranean **magma chamber**.[213] The relatively continuous, high rate of mostly basaltic volcanism from 11,950 to 9,375 years before the present, possibly assisted by increases in solar radiation or more credibly by gravitational stress due to Milanković cycles, was apparently sufficient to warm the ocean, and thus Earth, out of the last ice age.

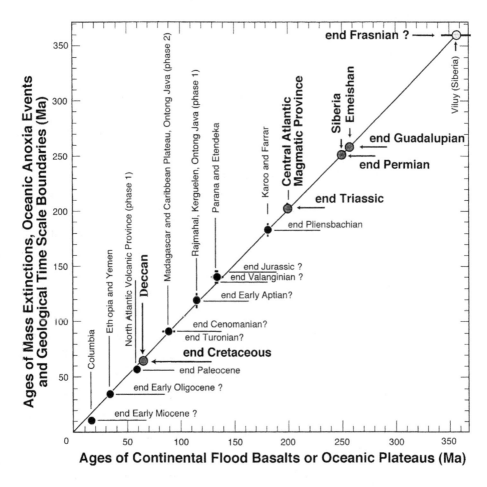

Figure 8.14 Flood basalts labeled vertically above the diagonal line and dated on the x-axis in megaanna (Ma), meaning millions of years before present, typically occur at the same time as mass extinctions dated on the y-axis and labeled horizontally below the diagonal line.

Volcanism (presumably basaltic) was similarly unusually active, primarily in Iceland, during the 13 Dansgaard-Oeschger sudden warming events between 46,000 and 11,600 years ago (see Figure 8.8).[214] During these sudden warmings, regional surface temperatures rose to interglacial levels within a decade or two and then decreased back to ice-age levels over many decades to centuries, most likely because volcanism waned before it could warm the deep ocean.

Approximately 56.1 million years ago, there was a brief period of extreme global warming as Greenland and Norway began drifting apart, forming the Greenland-Norwegian Sea (Figure 8.13).[215]

Suddenly, the amount of magma erupted at Earth's surface increased to rates greater than 3000 km³ per kilometer of rift per million years.[216] The extrusion of basalt may have lasted only 220,000 years, however.[217] Driven, presumably, by volcanic **halogen** (chlorine and bromine) emissions, global surface temperatures rose by 9 to 16°F (5 to 9°C) within a few thousand years. Sea surface temperatures near the North Pole increased to 73°F (23°C).[218] Southwest Pacific sea surface temperatures rose rapidly to 93°F (34°C), cooling back to 70°F (21°C) over seven million years.[219]

The Laki eruption, in 1783, extruded 3 mi³ (12.3 km³) of lava, which flowed over an area of 218 mi² (565 km²). Over the past 375 million years, however, (Figure 8.14) every 6 to 52 million years, 22 on average, there have been massive eruptions of flood basalts covering 0.08 to 3 million mi² (0.2 to 8 million km²). These have typically been contemporaneous with major mass extinctions.[220]

For example, approximately 250 million years ago, at the end of the Permian period, the Siberian Traps, comprising at least 0.7 mi³ (3 million km³) of basalt, were extruded in Siberia over an area of at least 2 million mi² (5 million km², equivalent to 62% of the contiguous 48 United States), possibly in less than 670,000 years.[221] Low-latitude surface seawater temperatures rose by 14°F (8°C),[222] nearly 3 times the rise in tropical Pacific sea surface temperatures at the end of the last glaciation. "Lethally hot temperatures exerted a direct control on extinction and recovery."[223] "Global warming and ozone depletion were the two main drivers for the end-Permian environmental crisis."[224] There was massive depletion of ozone.[225] "Prolonged exposure to enhanced UV radiation could account satisfactorily for a worldwide increase in land plant mutation" at this time.[226]

The most recent eruption of flood basalt was the Columbia River Basalts in Oregon and Washington States at approximately 16 million years ago, which produced an estimated volume of 50,500 mi³ (210,000 km³) of basalt.[227] A recurrence of even this smallest of flood basalt events would be a major, and quite possibly fatal, calamity for humankind.

Volcanism (presumably basaltic) was similarly unusually active, primar[ily] during the 13 Dansgaard-Oeschger sudden warming events between 46,000 a[nd] years ago (see Figure 8.8).[214] During these sudden warmings, regional surface temp[erature] rose to interglacial levels within a decade or two and then decreased back to ice-age l[evels] over many decades to centuries, most likely because volcanism waned before it could warm the deep ocean.

Approximately 56.1 million years ago, there was a brief period of extreme global warming as Greenland and Norway began drifting apart, forming the Greenland-Norwegian Sea (Figure 8.13).[215]

Suddenly, the amount of magma erupted at Earth's surface increased to rates greater than 3000 km³ per kilometer of rift per million years.[216] The extrusion of basalt may have lasted only 220,000 years, however.[217] Driven, presumably, by volcanic **halogen** (chlorine and bromine) emissions, global surface temperatures rose by 9 to 16°F (5 to 9°C) within a few thousand years. Sea surface temperatures near the North Pole increased to 73°F (23°C).[218] Southwest Pacific sea surface temperatures rose rapidly to 93°F (34°C), cooling back to 70°F (21°C) over seven million years.[219]

The Laki eruption, in 1783, extruded 3 mi³ (12.3 km³) of lava, which flowed over an area of 218 mi² (565 km²). Over the past 375 million years, however, (Figure 8.14) every 6 to 52 million years, 22 on average, there have been massive eruptions of flood basalts covering 0.08 to 3 million mi² (0.2 to 8 million km²). These have typically been contemporaneous with major mass extinctions.[220]

For example, approximately 250 million years ago, at the end of the Permian period, the Siberian Traps, comprising at least 0.7 mi³ (3 million km³) of basalt, were extruded in Siberia over an area of at least 2 million mi² (5 million km², equivalent to 62% of the contiguous 48 United States), possibly in less than 670,000 years.[221] Low-latitude surface seawater temperatures rose by 14°F (8°C),[222] nearly 3 times the rise in tropical Pacific sea surface temperatures at the end of the last glaciation. "Lethally hot temperatures exerted a direct control on extinction and recovery."[223] "Global warming and ozone depletion were the two main drivers for the end-Permian environmental crisis."[224] There was massive depletion of ozone.[225] "Prolonged exposure to enhanced UV radiation could account satisfactorily for a worldwide increase in land plant mutation" at this time.[226]

The most recent eruption of flood basalt was the Columbia River Basalts in Oregon and Washington States at approximately 16 million years ago, which produced an estimated volume of 50,500 mi³ (210,000 km³) of basalt.[227] A recurrence of even this smallest of flood basalt events would be a major, and quite possibly fatal, calamity for humankind.

How Do the Properties of Explosive and Effusive Volcanic Eruptions Compare?

The observed properties of historic and ancient explosive and effusive volcanic eruptions are summarized in Table 8.1. Basalt is a relatively dense, high-temperature, primitive magma type that rises directly from depths within Earth's **mantle** into the crust. When the crust is thin, as it is beneath oceanic islands such as Iceland and Hawaii, hot basalt has enough buoyancy to rise to form **sills** and small magma chambers at depths of a few miles and then to extrude out onto the land surface with little explosive action, forming instead lava fountains and "curtains of fire." Explosions are typically too small to eject substantial volcanic ash and gases into the stratosphere. Extrusions are typically very voluminous, accumulating over weeks to millennia.

Where the crust is thick, as it is beneath the continents, basalt magma only has enough buoyancy to rise to depths of 3 to 9 mi (5 to 15 km), where it remains hot enough to melt and assimilate crustal materials, thereby forming more evolved, lower-temperature, and less dense magmas that are richer in silicon dioxide than basalt. Over hundreds to hundreds of thousands of years, the density of the magma decreases, and the gas content increases, until some of the magma can rise to the surface, causing a large explosive eruption. Sometimes, so much magma is ejected in a giant eruption that the overlying rocks collapse into the magma chamber, forming a vast, crater-like depression known as a **caldera**. Explosive eruptions tend to last only hours to days, but an individual volcano may erupt repeatedly over intervals of centuries to hundreds of thousands of years.

These volcanoes do not have much direct effect on climate, but they do heat the ocean. Chlorine and bromine emissions from **submarine** volcanoes would be absorbed by the ocean, but massive releases of hydrogen sulfide from submarine volcanoes, common during **anoxic** (oxygen-poor) intervals of Earth history, would be able to enter the atmosphere and thus contribute to ozone depletion and warming.[232] It is possible for large basaltic eruptions to explode some volcanic gases and debris into the lower stratosphere, forming aerosols, but in contrast to explosive eruptions, this is typically a low percentage of the overall erupted material.

In summary, large explosive volcanic eruptions have been observed throughout human history to cause global cooling of up to 1°F (0.6°C) for up to 3 years and can increment Earth into ice ages when occurring every decade or so. Voluminous effusive eruptions of basalt lasting months to hundreds of thousands of years, however, have typically been contemporaneous with major global warming of many degrees throughout both human and geologic history. The balance between how much explosive volcanism and how much effusive volcanism occurs at any given time is determined by the motions

Table 8.1 Properties of historic and ancient explosive and effusive volcanic eruptions. DRE is short for Dense Rock Equivalent volume. Ma is million years ago. Ppmv is parts per million by volume. Gt is gigatons. Approximately 80% of all volcanism on Earth occurs along spreading rifts at mid-ocean ridges, most of which lie more than 1.2 mi (2 km) below sea-level.

	An Explosive Eruption	An Effusive Eruption	Extreme Explosive Eruption	Extreme Effusive Eruption
Volcano	Pinatubo (1991)	Laki (1783)	Yellowstone (2.2 Ma)	Siberian Traps (250 Ma)
Duration of longest eruptive phase	9 hours	8 days		
Duration of emissions	5 days	8 months		670,000+ years
Volume of eruptives (DRE)	5 km³			
Bulk volume of pyroclastic fallout	4 km³	0.4 km³		
Dominant type of mass flow	pyroclastic (ignimbrite)	basaltic lava	pyroclastic (ignimbrite)	basaltic lava
Bulk volume of mass flow	6 km³	15 km³	2,450 km³	3 million km³
Area of mass flows	400 km²	565 km²	15,500 km²	5 million km²
Eruption columns to	>35 km	>13 km		
SO2 emissions	17 Mt	122 Mt		7 Gt sulfur
H2O emissions	491-921 Mt	235 Mt		
Chlorine emissions	3-16 Mt	15 Mt		
Fluorine emissions		7 Mt		
Bromine emissions	>11-25 kt			
Maximum regional sulfur dioxide concentrations	300 ppbm	1000 ppbm		
Average effect on climate	global cooling -0.5°C for 3 years	warming		+8°C warming low-latitude sea surface
Warming of lower stratosphere	+3°C within 5 months			
Peak change in surface temperature	-0.7°C		-10°C for a few months	
Northern continental winter temperature	+3°C within 5 months			
Primary references	228	229	230	231

of Earth's tectonic plates. This balance has a dominant effect on climate because, although both types of volcanic eruptions deplete ozone, causing warming, only explosive volcanism causes cooling.

How Does Volcanism Deplete Ozone?

The longest continuous measurements of total column ozone (black line in Figure 8.15) have been made since 1927 at Arosa, Switzerland.[233] The dashed gray line with blue data markers shows, for 1964 to 2009, the annual mean area-weighted total ozone deviation from the 1964 to 1980 means for northern mid-latitudes (30°N to 60°N) scaled from -8%

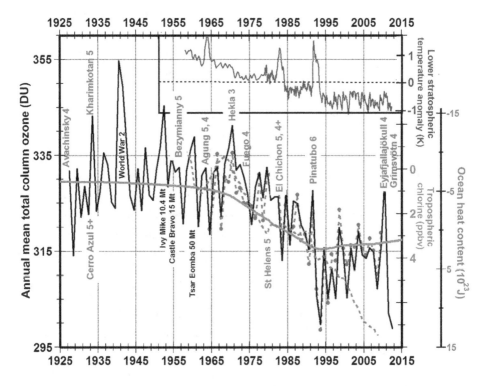

Figure 8.15 Mean annual total column ozone above Arosa, Switzerland (black line), anthropogenic tropospheric chlorine (green line), ocean heat content (dashed red line) and lower stratospheric temperature anomaly (purple line). Note that the y axes of the green and dashed red lines are inverted. The dashed gray line with blue data markers shows, for 1964 to 2009, the annual mean area-weighted total ozone deviation from the 1964 to 1980 means for northern mid-latitudes.

at the bottom of the figure to 10% at the top.[234] Years of increasing or decreasing ozone are nearly identical for Arosa and for this area-weighted mean, with small differences in amplitude. Thus, the Arosa data provide a reasonable approximation for annual mean total column ozone throughout northern mid-latitudes since 1927.

Ozone at Arosa averaged 331 **Dobson units** (DU) until 1974, fell 9.4% to 300 DU by 1993, and began generally rising again until 2011. The long-term decrease in ozone has been reasonably attributed to the chlorine-catalyzed destruction of ozone due to an increase in the concentration of anthropogenic tropospheric chlorine (green line, y axis inverted). In recognition of this very serious problem, the **Montreal Protocol**, mentioned previously, was ratified, beginning in 1987, leading to the phasing out of the production of **chlorofluorocarbons** (CFCs) and to a consequent decrease in tropospheric chlorine, which began in 1993. Long-term chlorine concentrations are expected to return by 2040 to levels that were prevalent before the late-1970s.[235]

The dashed red line shows how ocean heat content increased (scale inverted) as ozone became depleted and continues to increase as long as ozone remains depleted relative to pre-1965 levels, as expected.

The lowest levels of annual mean total column ozone were observed in 1992 and 1993 following the 1991 eruption of Mt. Pinatubo, the largest explosive eruption since that of Mt. Katmai in 1912. Similarly low ozone levels were also recorded in 2011 and 2012 following the eruption of Eyjafjallajökull in Iceland in 2010, one of the larger effusive eruptions in the past century. The size of an explosive volcanic eruption is typically measured using the logarithmic **Volcanic Explosivity Index** (VEI), based primarily on the volume of tephra (volcanic debris) erupted and the maximum height of the eruption column. The size of effusive volcanic eruptions is best measured using the volume of magma extruded.

Volcanoes labeled in red in Figure 8.15 include all very large explosive eruptions with VEI ≥5, some VEI 3 and 4 effusive eruptions in Iceland, and two explosive VEI 4 eruptions from volcanoes elsewhere that typically included substantial lava flows. All volcanic eruptions shown were followed a year later by a substantial decrease in annual mean total column ozone except the 1980 eruption in Washington state of Mt. St. Helens, which was a phreatic explosion—an unusual blast of steam from a water-rich magma triggered by a landslide. Most other small volcanic eruptions since 1927 appear to have caused similar changes in ozone, but the signal-to-noise ratios are too small to draw reasonable conclusions.

There is an increase in ozone in the year of most of these eruptions compared to the previous year, as discussed in greater detail below. There are also large peaks in annual mean total column ozone during years containing the three largest atmospheric nuclear tests, labeled in black with yield in megatons. The largest short-term peak in ozone was in 1940 and 1941 during the major territorial conquests of World War II. Any causal relationship is unclear. The most consistent short-term changes in ozone are an apparent increase during the year of a volcanic eruption followed by a much larger depletion during the next few years.

Pinatubo (VEI 6) erupted in the Philippines on June 15, 1991, followed closely by the eruption of Cerro Hudson (VEI 5+) in southern Chile on August 12, 1991. Annual mean ozone increased 3.9% in 1991, primarily between February 19 and 26. By 1993, however, annual mean ozone dropped 8.5% to 300 DU, the lowest level ever recorded as of that time.[236] Total column ozone was 11 to 17% below preceding years throughout Canada, with a peak loss of 30% at approximately 10 mi (16 km) altitude.[237] On average, total ozone decreased 8% in Europe, 5 to 6% in North America, Russia, and Asia, but less than 2% in the tropics. Following the 1982 eruptions of El Chichón in Mexico (VEI 5 and 4+), total ozone similarly decreased 5% in Europe, 3% in North America and Russia, and less than 1% in the tropics. Following the 1963 eruptions of Agung in Bali (VEI 5 and 4), total ozone fell 5% in Europe and Asia, 2% in North America, and less than 1% in the tropics.

An even larger ozone anomaly in 2010 is associated with the 100-times less-explosive basaltic effusive eruption of Eyjafjallajökull in South Iceland (VEI 4). A slightly larger eruption of Grímsvötn (VEI 4), 85 mi (140 km) northeast of Eyjafjallajökull, occurred in May 2011, compounding the amount of ozone depletion during 2011 and 2012. The amplitudes of these short-term ozone anomalies since 1990 are larger than the amplitudes of earlier volcanic anomalies before the global rise in anthropogenic tropospheric chlorine (green line, y-axis inverted). Similar anomalies appear to be associated with the eruption of Hekla in Iceland (1970, VEI 3).

Following the eruption of Pinatubo, the lower troposphere warmed by up to 5.4°F (3°C) during the winter throughout the northern parts of northern continents,[238] the areas with greater depletion of ozone. Related major changes in atmospheric chemistry are well documented by a 45% drop in total column nitrogen dioxide above Switzerland beginning five months after the Pinatubo eruption and returning to normal within two years,[239] a 40% decrease in nitrogen dioxide observed above New Zealand,[240] and substantial increases in nitric acid.[241]

The observed ozone depletion (Figure 8.15) was accompanied by cooling of the stratosphere (purple line) occurring mostly "as two downward 'steps' coincident with the cessation of transient warming after the major volcanic eruptions of El Chichón and Mount Pinatubo" and a similar downward step following the 1963 eruptions of Agung volcano.[242] This cooling reflects a decline in the heat-generating absorption of UV-B solar radiation by the thinned ozone layer in the lower stratosphere. Aside from these transient effects, however, the overall downward trend of the stratospheric temperature curve faithfully reflects the ozone depletion by chlorine from CFCs from the early 1960s through 1993, when the downward trends in all three curves were abruptly terminated by the Montreal Protocol.

Ozone depletion following volcanic eruptions has traditionally been explained by new aerosols formed in the lower stratosphere providing substantial new surfaces on which heterogeneous chemical reactions can form ozone-destroying chlorine at cold temperatures in a **polar vortex**. More significant, however, is the fact that water vapor, the most voluminous gas erupted from volcanoes, is a primary source of hydroxyl radicals that **catalytically** destroy ozone in the stratosphere.[243] Volcanoes also erupt megatons of halogens, primarily as hydrogen chloride and hydrogen bromide, and, as noted previously, only one halogen molecule can catalytically destroy more than 100,000 molecules of ozone.[244] During explosive eruptions, many of these halogens appear to be removed immediately from the eruptive cloud in condensed, supercooled water.[245] Eruptions from drier, effusive basaltic volcanoes, however, such as Eyjafjallajökull and Grímsvötn, do not form significant eruption columns that remove halogens and create aerosols in the stratosphere. Furthermore, even though they emit less water vapor than explosive volcanoes, they typically involve ten to 100 times more volatile components other than water vapor per unit volume of magma than do explosive eruptions.[246]

Ozone depletion is substantial within the plumes of erupting volcanoes. Detailed observations imply "the most likely cause for the observed rapid and sustained ozone loss to be catalytic reactions with halogen, mainly bromine, radicals."[247] Recent observations have shown that even the plumes of quiescently degassing volcanoes are chemically very active, containing halogens that modeling shows cause ozone depletion upon reaching the stratosphere.[248] Between erupting abundant halogens and injecting large amounts of water into the stratosphere, volcanoes appear to deplete ozone along many more chemical paths than those attributed to anthropogenic chlorofluorocarbons.

Volcanic eruptions are typically followed a year later by an approximately 6% depletion of ozone averaged throughout the year. How do these short-term effects of volcanism compare to the longer-term effects of anthropogenic chlorofluorocarbons? The green line for anthropogenic tropospheric chlorine in Figure 8.15 is inverted and has been scaled so that its increase from 1965 to 1993 has approximately the same rate of change as the corresponding long-term decrease in ozone, as expected by current theory. This visual fit suggests that depletion of ozone following the Pinatubo eruption (approximately 20 Dobson units, DU) was twice as large as the depletion due to anthropogenic chlorofluorocarbons since 1960 (approximately 10 DU), and that it takes more than a decade after an eruption for ozone concentrations to return to pre-eruption levels.

In summary, numerous basaltic effusive eruptions, such as Eyjafjallajökull and Grímsvötn, as well as quiescently degassing, non-eruptive volcanoes deplete ozone, causing global warming. Large, but less frequent, explosive volcanic eruptions also deplete ozone, causing warming at Earth's surface especially during the first winter, but also eject megatons of water and sulfur dioxide into the lower stratosphere, forming, over many months, sulfur dioxide aerosols that reflect, scatter, and absorb solar radiation, causing cooling at Earth's surface of up to 1°F (0.6°C) over a period of up to three years.

How Common Is Abrupt Climate Change?

According to the $\delta^{18}O$ **proxy** for **paleotemperatures** in ice cores from Greenland,[249] local temperatures rose 18 to 29°F (10 to 16°C) in less than 40 years[250] at least 25 times, and perhaps more than 70 times, between 115,000 and 15,000 years before 2000 AD. Temperatures then returned to ice-age conditions more slowly, usually within centuries (Figure 8.16).[251]

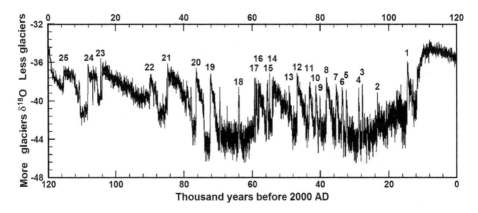

Figure 8.16 The $\delta^{18}O$ proxy for paleotemperatures in ice cores from GRIP cores showing the 25 basic Dansgaard-Oeschger sudden warmings. Data from the North Greenland Ice Core Project (NGRIP) holes 1 and 2.

These abrupt warmings might even have happened within years. For example, "the deuterium excess, a proxy for Greenland precipitation moisture source, switches mode within 1 to 3 years over these transitions,"[252] documenting very rapid changes. These sudden warmings, known as Dansgaard-Oeschger events, occurred on average every 1440 years, but at intervals that were more random than periodic. Broecker and Denton[253] suggest that such "rapidity of glacial terminations" and "hemispheric synchronization" can only be explained by major reorganizations of the ocean-atmosphere system most likely triggered by sudden influxes of fresh water that changed the North Atlantic circulation. The extensive literature over the past 25 years provides considerable detail from this perspective but fails to provide convincing evidence for how large volumes of fresh water from glacial melt can precede and therefore cause the warming. Since explosive volcanoes are well known to form stratospheric aerosols that cause cooling, most scientists have not considered the compelling evidence that effusive, basaltic volcanism might cause warming and that the Dansgaard-Oeschger events might therefore have been driven by this style of volcanism, which is well documented to have occurred under ice during these times.

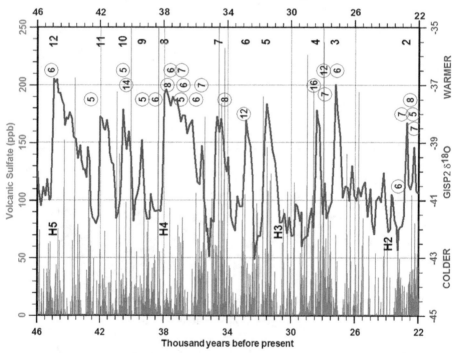

Figure 8.17 Dansgaard-Oeschger sudden warming events (numbers along the top) all correspond to times of continuous volcanism. Red bars show the amount of sulfate in individual layers of ice in the GISP2 borehole. The purple line shows the $\delta^{18}O$ proxy for temperature adjusted for gas age. Numbers in blue circles show the number of contiguous layers containing sulfate anomalies at the time plotted. H2 to H5 are Heinrich events when large numbers of icebergs suddenly appeared in the northern Atlantic Ocean.

The red bars in Figure 8.17 show volcanic sulfate anomalies from 46 to 22 thousand years ago from the GISP2 borehole under Summit Camp, Greenland. The numbers in the blue circles indicate the number of **contiguous layers** containing volcanic sulfate (only clusters of 5 or more such layers are shown). Each sampled layer in this section of the ice core represents an average of 11 years. The purple line represents temperature, based on the $\delta^{18}O$ proxy. Numbers 2 thru 12 across the top of the figure represent Dansgaard-Oeschger events. All warming phases except event 5 at 31.5 thousand years ago are contemporaneous with volcanic sulfate in 5 or more contiguous layers. Event 5 is contemporaneous with much higher volcanic sulfate (795 ppb) than usual between 31,632 and 31,203 years before the present, just not in more than 4 contiguous layers throughout the sequence.

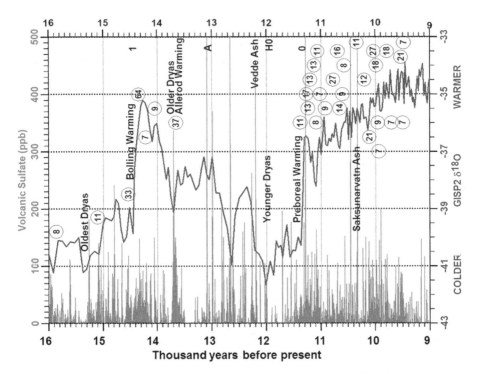

Figure 8.18 Dansgaard-Oeschger sudden warming events (0, A, and 1 along the top) all correspond to times of continuous volcanism. Red bars show the amount of sulfate in individual layers of ice in the GISP2 borehole. The purple line shows the $\delta^{18}O$ proxy for temperature adjusted for gas age. Numbers in blue circles show the number of contiguous layers containing sulfate anomalies at the time plotted. H0 is a Heinrich event, in which large numbers of icebergs suddenly appeared in the northern Atlantic Ocean.

Figure 8.18 shows similar data from 16 to 9 thousand years ago. The association in time of volcanic sulfate and warming is particularly clear during the **Bølling** and Preboreal warmings, as are major decreases in sulfate during the colder periods: the Oldest Dryas, the Older Dryas, and the Younger Dryas.

The ocean, whose surface covers 71% of Earth's surface, provides the primary heat capacity in the climate system.[254] The heat capacity of the whole atmosphere is equal to that of approximately the top 3.2 meters of the ocean, yet the average depth of the ocean is 3686 meters. Thus, true global warming entails not only warming the atmosphere, but the deep ocean as well. Equatorial Pacific sea surface temperatures during the last glacial maximum were approximately 5.4°F (3°C) lower than today.[255]

All major warming events in the past 46,000 years are contemporaneous with high rates of volcanism. Even short-term warming appears contemporaneous with increased volcanism when plotting the data on time scales sufficient to study individual layers. The data are most numerous and convincing since the last glacial maximum and decrease in number with age back to the lowest layers in the ice core (110,400 years before the present). Generally, when the amounts of volcanic sulfate increase rapidly for many years, global warming occurs. When this warming continues for decades to centuries, the ocean surface water is warmed. When the volcanism slows and stops, the deep thermal reservoir of the ocean cools Earth back into an ice age. When volcanism lasted a couple of thousand years during the pre-Boreal warming, the effect was sufficient to heat the ocean's deep thermal reservoir enough to warm Earth out of the ice age.

Why Does Ozone Peak in the Year in Which Volcanoes Erupt?

The ozone increase in 2010 began just before the onset of eruptions at Iceland's Eyjafjallajökull volcano between February 19 and February 26. This was approximately 4 weeks before the first effusive flank eruption of basalt from March 20 to April 12 and approximately 7 weeks before the main explosive eruption of **trachyandesite** on April 14. Total ozone northeast of Iceland increased on February 19 to more than 550 DU above a background concentration of approximately 325 DU, an increase of about 70% (Figure 8.19).[256]

Global total ozone over the period from February 21 to February 28 increased 45% above the mean level from 1978 through 1988. The sudden onset of these emissions is shown clearly in an animation of daily ozone maps from December 1, 2009 to April 30, 2010.[257]

At Eyjafjallajökull, seismic activity and deformation began in December 2009, "explained by a single horizontal sill inflating at a depth of 2.5 to 3.7 mi (4.0 to 5.9 km) under the south-eastern flank of the volcano" (Figure 8.20).[258]

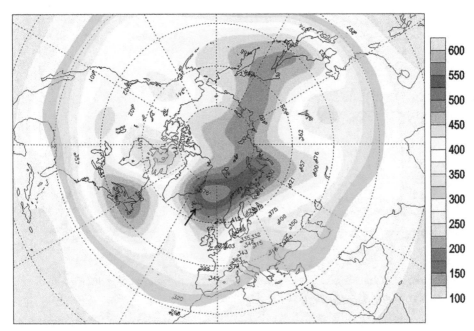

Figure 8.19 Total column ozone northeast of Iceland increased to more than 550 DU on February 19, 2010, over a background of ~325 DU, an increase of ~70%. Arrow shows location of the volcano Eyjafjallajökull.

Deformation increased exponentially in February, suggesting a major change in pressure conditions within the system by March 4. Thus, a substantial release of gas from the top of the magma body is highly likely to have occurred in late February as the roof of the intrusion fractured to the surface.

Where could this pre-eruption ozone be coming from? It is conceivable that **partial pressure** of oxygen in magma is sufficient to concentrate oxygen gas in the very high temperature (high energy) environment at the top of a magma body, where ozone could be formed and then released when fractures first propagate to the surface. Oxygen and ozone are not likely to occur in volcanic gases, however, because magma typically has a reduced oxidation state.

The "ozone" signal could possibly be confused with sulfur dioxide, a primary volcanic gas emitted from high-temperature **fumaroles**, that absorbs ultraviolet solar radiation strongly at frequencies similar to those at which ozone absorbs, and therefore sulfur dioxide is known to affect ozone measurements in urban areas such as Uccle, just outside

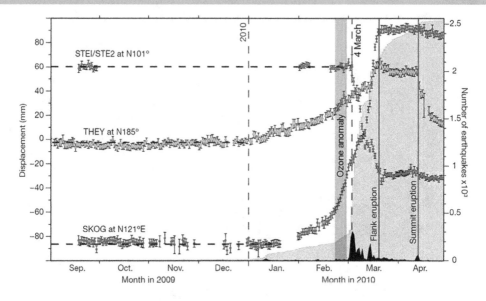

Figure 8.20 Ozone emission (red region) occurred just as magma began breaking to the surface at Eyjafjallajökull.

of Brussels, Belgium.[259] Sulfur dioxide is colorless, so that its release from high-temperature fumaroles in the vicinity of the effusive eruption that began on March 20 would have gone unnoticed if substantial water vapor were not included. There were no instruments in the vicinity to detect either sulfur dioxide or ozone, but satellite data are normally sufficient to distinguish ozone from sulfur dioxide. Local farmers did note unusual melting of snow near high temperature fumaroles "months" before the eruption of Hekla in south Iceland on May 5, 1970.[260]

Another possibility is that high-temperature (high-energy) gases from basaltic magma (2400 to 2550°F, 1300 to 1400°C)[261] may interact in some way with water released from the magma, with ground water, or even with atmospheric water vapor, to split oxygen atoms and form ozone. Magmatic high-temperature gases might also interact with volatile organic compounds or nitrogen compounds or gases to form ozone catalytically.

An even more likely possibility is that the ozone was generated by rock fracture as the gases at the top of the magma broke to the surface. Crushing and grinding small samples of igneous and metamorphic rocks at atmospheric pressure in the laboratory generated up to 10 **parts per million** ozone, apparently "formed by **exoelectrons** emitted by high electric fields resulting from charge separation during fracture."[262]

A similar increase in ozone was observed north of Mt. Pinatubo, in the Philippines, as the volcano showed signs of reawakening that included a group of felt earthquakes on

March 15, 1991, the first steam explosions on April 2, the first eruption on June 12, and the main eruption on June 15. Ozone anomalies greater than 550 DU, a 40% increase, occurred from February 20 to 22, 1991, well north of the volcano.[263]

Maps of ozone distribution in each hemisphere are produced daily. Consulting these maps when a volcano begins to show signs of renewed activity might provide a way to predict whether and when an eruption might occur. There is considerable need for research here.

Summary of How Volcanic Eruptions Cool and Warm Earth

Extensive observations, discussed in this chapter, suggest strongly that increases in the rate of explosive volcanic eruptions cool Earth even into ice ages while increases in the duration of effusive basaltic volcanic eruptions can warm Earth substantially. The processes involved are summarized in Figure 8.21.

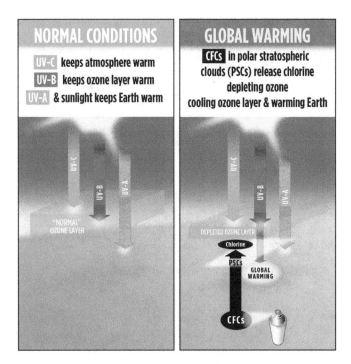

Figure 8.21 a) Under conditions normal before 1965, ultraviolet-C (UV-C) warmed the upper atmosphere, UV-B primarily warmed the ozone layer, and UV-A and visible light warmed Earth. b) CFCs, when they rise to the level of very cold polar stratospheric clouds (PSCs), release chlorine that depletes ozone, causing more UV-B than usual to reach Earth's surface, thus cooling the ozone layer and warming Earth.

In Panel a, under conditions that were normal before 1965, ultraviolet-C (UV-C) radiation warmed the upper atmosphere, UV-B radiation primarily warmed the ozone layer, and UV-A radiation and visible light warmed Earth. In Panel b, CFCs, when they rise to the level of very cold **polar stratospheric clouds** (PSCs), release chlorine that depletes ozone catalytically, allowing more UV-B radiation than usual to reach Earth's surface, reducing the warming of the ozone layer and increasing the warming of Earth's surface. In Panel c, effusive volcanoes emit chlorine and bromine, which deplete ozone, leading to global warming. In Panel d, explosive volcanoes similarly deplete ozone, but they also eject megatons of water and sulfur dioxide into the lower stratosphere, forming globe-encircling sulfur dioxide aerosols that become extensive enough—and whose particle sizes grow large enough—to reflect and scatter sunlight, especially UV-B and UV-A radiation, causing net global cooling.

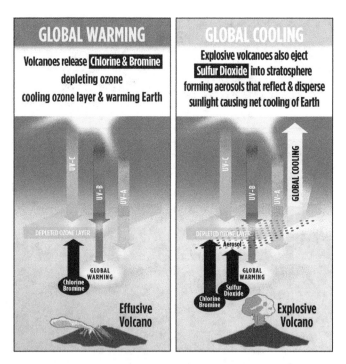

Figure 8.21 c) Effusive volcanoes emit chlorine and bromine, which deplete ozone, leading to global warming. d) Explosive volcanoes similarly deplete ozone, but also eject megatons of water and sulfur dioxide into the lower stratosphere, forming globe-encircling aerosols that become extensive enough to reflect and scatter sunlight, causing net global cooling.

CHAPTER 9

HOW DO VOLCANIC ERUPTIONS AFFECT WEATHER?

"Science is the outcome of being prepared to live without certainty and therefore a mark of maturity. It embraces doubt and loose ends."
    —A. C. Grayling, 2009

Weather is the primary result of the flow of heat from the tropics to the poles, and the distribution of this heat is influenced and augmented by the formation and **photodissociation** of ozone. A fundamental conclusion of the science discussed in this book is that volcanic eruptions affect the **ozone layer**. Now, let's explore how spatial and temporal variations in ozone concentrations can affect weather.

The ozone layer was first discovered in 1913 by Charles Fabry and Henri Buisson, but its properties were explored in far greater detail by Gordon Dobson, a physicist and meteorologist. Dobson noticed, while studying meteorites passing through the atmosphere, that the **temperature** profile of the **tropopause**, 11 mi (17 km) above the tropics and 5.6 mi (9 km) above the poles, was not constant, as was believed at the time. (Go back and look at Figure 6.6 on page 86 for a more recent observation of a substantial change in this boundary over just a few hours.) In order to explain this phenomenon, Dobson reasoned that ultraviolet solar **radiation** must be heating ozone in the lower **stratosphere**.

Dobson built the first spectrophotometers to measure total column ozone looking up from Earth's surface.[264] Accordingly, concentrations of ozone are now reported in **Dobson Units** (DU). One DU defines a layer of gas that would be ten micrometers thick at standard temperature and pressure. In other words, 300 DU of ozone brought down to Earth's surface at zero degrees Centigrade and standard sea level pressure would occupy a layer only 3 millimeters thick.

In Chapter 6, we discovered that the lifetime of a single molecule of ozone is only about 8.3 days, on average, and that this is determined by the Chapman Cycle, driven by **ultraviolet radiation** from Sun, which maintains the ozone layer by continually creating and destroying ozone in the lower stratosphere. Heat is released both when ozone is created through photodissociation of oxygen and when ozone itself is dissociated. Thus, the concentration of ozone at any particular point in the atmosphere indicates how much heat is being generated in that location, but that estimate is spread out by how far each molecule can travel during its lifetime.

What Is the Link Between Ozone and Weather?

In 1950, Richard Reed,[265] a senior professor of Atmospheric Sciences at the University of Washington, wrote: "From the time of Dobson's early measurements [in the 1920s], it has been known that the total ozone amount undergoes large day-to-day fluctuations. Data from stations in many parts of the world have revealed that these fluctuations are everywhere greatest in winter and spring, while in any one season they are greater at higher than at lower latitudes. ... However, the day-to-day changes of total amount are not only of interest because of their amplitude—which, incidentally, is nearly as large as the amplitude of the seasonal variation—but also because of the relationships which have been established between them and various meteorological quantities."[266]

As mentioned previously, Dobson[267] showed that maximum positive daily deviations of ozone values from monthly means are generally found to the rear of surface low-pressure areas, while maximum negative deviations are found to the rear of surface highs. On the basis of more extensive measurements, Dobson[268] found later that for many occluded fronts, when a cold front overtakes a warm front and forces warm air aloft, the maximum positive deviations occur directly over the surface low rather than to the rear of it. Tonsberg and Langlo-Olsen[269] showed that a surface cold front or occluded front is the dividing line between positive and negative departures from normal. Meetham and Dobson[270] found that the total amount of ozone correlates positively with temperature in the lower stratosphere and negatively with density and height of the tropopause. With the advent of satellite systems, it has become possible to observe these variations and their distribution with increasing precision.[271]

Dobson and others had little evidence for any physical or chemical processes that might change ozone concentrations directly other than the dynamical effects of weather systems scattering and concentrating ozone. Had they understood the thermal effects of the Chapman cycle, they would have had a better understanding of the reasons for these variations, but at that time, the cycle and its effects were still unknown.

Staehelin and others[272] summarize that "extra tropical total ozone values vary in concert with a number of dynamical parameters such as the free tropospheric temperature, the lower stratospheric temperature, the geopotential height, the potential vorticity of the lowermost stratosphere, the tropopause height, and the tropopause pressure."

Similarly Shephard[273] concludes that "chemical ozone loss is controlled by atmospheric dynamics. Dynamics explains why it is that ozone loss is much more severe in the Antarctic springtime than it is in the Arctic springtime. Dynamics also explains why ozone loss exhibits year-to-year variability even while the **halogen** loading evolves more smoothly and why this variability is largest in the northern hemisphere. Moreover, dynamical variability also leads to changes in ozone abundance through changes in transport. Observed changes in ozone thus result from both dynamical and chemical effects, in a non-additive manner."

Ozone, in the presence of sufficiently energetic ultraviolet radiation, causes warming, no matter what the shape and extent of the ozone mass. Clear evidence exists for the role of ozone depletion in reducing lower stratospheric temperatures, increasing upper tropospheric temperatures, and allowing more high-energy solar ultraviolet radiation to reach Earth's surface, causing warming. As noted in Chapter 8, clear evidence also exists for rapid changes in ozone concentrations related to volcanism, even from small eruptions occurring many times per year. Together, these observations suggest that ozone variations may play a much more active role in atmospheric dynamics than currently assumed, and vice versa. Modeling these complex effects in the short-term, although exceedingly challenging, could improve weather forecasting substantially; modeling them in the long-term would be even more difficult, but could theoretically improve our understanding of global climate.

In any case, ozone depletion theory provides a much stronger link than greenhouse warming theory between short-term changes in weather and the long-term changes in weather known as climate change. The take-away message from all this is that ozone amounts and distribution respond to (and contribute to) atmospheric dynamics as well as responding to solar UV radiation and to chlorine from CFCs and **volcanoes**.

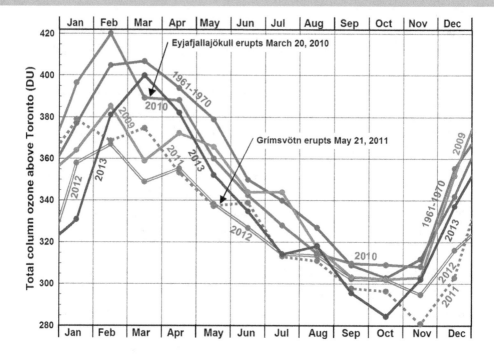

Figure 9.1 Ozone concentrations above Toronto, Canada, reached a minimum in 2012, when minimum temperatures reached a maximum (Figure 9.2). Total column ozone per month in November, 2011 (dotted line), was 12% below the average for Novembers in 1961 through 1970 (upper line) and has remained unusually low throughout 2012.

Warming and Drought in Toronto During 2012 and 2013

The blue line in Figure 9.1 shows the average monthly total column ozone measured above Toronto, Canada, for the years 1961 through 1970, during which time **anthropogenic** tropospheric chlorine had only increased 16% from levels in 1925 towards peak levels in 1993. The green line shows average monthly column ozone for 2009, after anthropogenic chlorine had decreased by 9% from its peak atmospheric concentration. Note that while ozone concentrations are highest between December and May, ozone depletion is also greatest between December and May, the months of the greatest warming throughout the northern hemisphere.[274] The mean change in radiation between the 1960s and 2009 caused by ozone depletion is directly proportional to the change in area between the blue and green lines.

The solid red line includes both the increase in ozone during February preceding the first eruption of Eyjafjallajökull in March 2010, discussed in the last chapter, and subsequent ozone depletion. The dotted red line shows total ozone in 2011, including significant depletion due to Eyjafjallajökull and relatively minor increases and decreases related to the smaller eruption of Grímsvötn. The double red line shows ozone levels in 2012; the purple line shows levels for 2013, indicating rapid recovery following the effects of the eruptions. Note that from November 2011 through February 2012, the times when monthly maximum temperature records have been set throughout central North America,[275] ozone has been depleted by as much as 14% below mean values in the 1960s.

The thin, dashed, black line in Figure 9.2 shows the average values of total column ozone above Toronto from December through April for each year from 1965 to 2012; the thin, dashed, red line (y-axis inverted) shows the average values of monthly mean minimum temperatures from December through April for the same time interval. The thick, solid

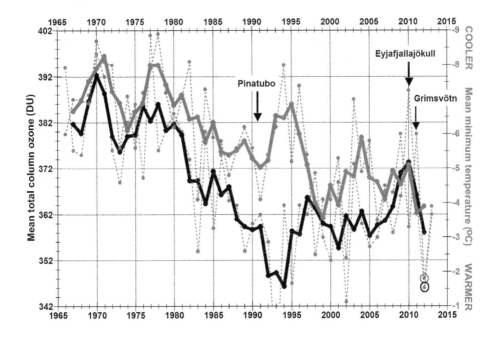

Figure 9.2 Ozone concentrations above Toronto, Canada, reached a minimum in 2012 (Figure 9.1), when minimum temperatures reached a maximum. When mean total column ozone measured during the months of December through April (lower lines) decreased, mean minimum temperature (Environment Canada, 2014) for the same months typically warmed (upper lines, y-axis inverted), except following the eruption of Pinatubo in 1991. The dashed lines show annual means; the solid lines are smoothed using a 3-point centered running mean.

lines show the same data after being smoothed with a 3-month symmetric running mean. The two curves show remarkably similar trends, except from 1992 to 1995, when **aerosols** in the lower stratosphere following the June 1991 eruption of Pinatubo decreased radiation from Sun by as much as 2.7 W m^{-2} during August and September,[276] decaying exponentially to negligible values by 1995. Note the extremely low ozone and high temperature in early 2012 (circled data points, lower right part of the image).

These two data points suggest that depletion of ozone due to the eruptions of Eyjafjallajökull in 2010 and Grímsvötn in 2011 supplemented anthropogenic depletion, leading to extreme ozone depletion and to the extreme warm temperatures and drought observed throughout the Great Plains of North America during late 2011 and 2012,[277] and to the highest sea surface temperatures ever recorded on the continental shelf off the northeastern United States during the first half of 2012.[278] A warmer North Atlantic Ocean was easier to evaporate, which most likely led to the increased precipitation and major floods in Great Britain and Ireland and elsewhere in Europe during 2012 and 2013.[279]

On July 12, 2012, there was an extreme melt event across 98.6% of the Greenland Ice Sheet.[280] Such melt events at high elevations in Greenland are extremely rare, last seen in 1889 and before that around 1200.[281] The 1889 melting was coincident with a major El Niño[282] and the onset of major drought in 1889 from Mexico to Saskatchewan.[283] These events followed the 1883 eruption of Krakatau (VEI 6), the 1886 eruption of Okataina (VEI 5), and several VEI 4 eruptions. These observations raise the possibility that the major El Niño in 1998 and associated warming might have been related to major ozone depletion following the eruptions of Pinatubo (VEI 6) and Cerro Hudson (VEI 5+) in 1991, after the cooling effects of the stratospheric aerosols had ended. These effects could then have been compounded by three VEI 4 **explosive** eruptions that occurred between 1992 and 1994.

The drought of 2012 approached the intensity of the great Dust Bowl droughts of 1934 and 1936,[284] which occurred after a highly unusual sequence of seven VEI 4 and 5 eruptions from 1931 through 1933 in Indonesia, Japan, the Kurile Islands, Kamchatka, Alaska, Guatemala, and Chile,[285] providing at least a partial explanation for the well-known warming of mean northern hemispheric surface temperatures during the 1930s, followed by cooling in the 1940s.[286]

What Caused the "Weird" Weather of 2014 and 2015?

On August 16, 2014, a swarm of earthquakes began under Bárðarbunga volcano beneath the northwestern part of the Vatnajökull ice sheet in eastern Iceland. More than 1600 earthquakes were recorded within 48 hours, only three of which were as large as magnitude 3. Most occurred at depths between 3.1 and 6.2 mi (5 and 10 km).[287] There may have been a volcanic eruption under the ice, but the first observed eruption began early on August 29 at the Holuhraun lava field just north of the ice sheet. This eruption continued until February 28, 2015, extruding 0.34 mi^3 (1.4 km^3) of **basaltic magma** over an area of 33 mi^2 (85 km^2),[288] the largest lava flow in Iceland since 1783, when the 8.8 times bigger eruption of Laki volcano,[289] just southwest of Vatnajökull, extruded 3 mi^3 (12.3 km^3) of basalt over an area of 218 mi^2 (565 km^2) (Figure 9.3).

Figure 9.3 Eruption of Bárðarbunga in the Holuhraun lava field on September 4, 2014, between 14:30 and 15:00. Photo by Peter Hartree (en.wikipedia.org/wiki/2014–15_eruption_of_Bárðarbunga).

Based on the discussion in the last chapter, it seems likely that the Holuhraun lava field north of Bárðarbunga did deplete ozone, causing cooler lower stratospheric temperatures and warmer temperatures close to Earth. We need to collect data throughout 2015 before this eruption can be added to Figure 8.15 on page 119. As discussed in chapter 6, increased ozone depletion causes the **polar vortex** to become stronger, colder, and more persistent (Figure 9.4).[290]

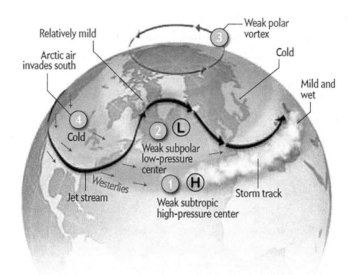

Figure 9.4 Changes in ozone concentrations effect changes in the polar vortex and jet stream, affecting changes in the distribution of cold Arctic air, as well as high and low pressure systems.

Changes in ozone concentrations cause changes in the shape and extent of the **polar jet streams** that can cause changes in the latitudes where precipitation occurs and changes in regions where excessive Arctic cold dips south into the eastern United States, Northern Europe, and Russia.[291]

It is going to take considerable research by many different people with many different interests to unravel all the connections between temporal and regional changes in ozone concentrations and temporal and regional changes in weather. Luckily, due in part to the **Montreal Protocol** on Substances that Deplete the Ozone Layer, considerable data regarding ozone are readily available.[292]

CHAPTER 10

WHY DOES THE GREENHOUSE EFFECT APPEAR NOT TO BE CORRECT?

"What gets us into trouble is not what we don't know. It's what we know for sure that just ain't so."

—Attributed to **Mark Twain**

The Glossary[293] of the 2013 report by the Intergovernmental Panel on Climate Change (IPCC) defines the greenhouse effect to be "*the infrared radiative effect of all infrared-absorbing constituents in the atmosphere. Greenhouse gases, clouds, and (to a small extent) aerosols absorb terrestrial radiation emitted by the Earth's surface and elsewhere in the atmosphere. These substances emit infrared radiation in all directions, but, everything else being equal, the net amount emitted to space is normally less than would have been emitted in the absence of these absorbers because of the decline of temperature with altitude in the troposphere and the consequent weakening of emission. An increase in the concentration of greenhouse gases increases the magnitude of this effect; the difference is sometimes called the enhanced greenhouse effect. The change in a greenhouse gas concentration because of anthropogenic emissions contributes to an instantaneous radiative forcing. Surface temperature and troposphere warm in response to this forcing, gradually restoring the radiative balance at the top of the atmosphere."*

Can Radiation From a Thermal Body Actually Warm the Same Body?

The fundamental assumption regarding the greenhouse effect is that **infrared** energy, radiated by Earth and absorbed by **greenhouse gases**, such as water vapor, **carbon dioxide**, and methane, ends up making Earth warmer. While absorption by these gases is clearly observed and well documented, the idea that infrared **radiation** originating from Earth can actually warm Earth is physically impossible. It is the thermal equivalent of a perpetual motion machine. Higher frequency and amplitude radiation from a warmer source than Earth must be added in order to make Earth warmer, just as energy must be added to any machine in order to overcome operational energy loss. A thermal body cannot spontaneously make itself warmer, as that would violate the second law of **thermodynamics**, which states that heat can only transfer from a warmer object to a cooler one—a fundamental principle in Nature to which no exception has ever been observed or recorded. If objects could actually heat themselves with their own radiation, that process would automatically escalate until the objects melted or vaporized, which clearly does not happen.

The zeroth law of thermodynamics states, in effect, that heat does not flow between bodies at the same temperature. Radiation from body A can warm body B only to a **temperature** that is as warm as body A. If body B is warmer than body A, then radiation from body B will flow to body A, warming it to a temperature that can be no higher than that of body B. If body A and body B have the same temperature, no heat can flow either way between them. Similarly, if you were to surround body A with a perfect reflector of the thermal radiation emitted from body A, no heat will flow back to body A because body A and the radiation from body A are at the same temperature. The only way to warm Earth is by radiation from a warmer body, such as Sun.

Nonetheless, it is a common assumption, central to the greenhouse effect, that radiation from a body of matter, such as Earth, will cause the body to get warmer if it is returned to that body. Where did this assumption originate? It actually derives from another assumption, "that energy comes *in little blobs of a definite amount*,"[294] and that you can take thermal energy from different places and simply add each of these energies together to get more thermal energy. Richard Feynman, one of the most noted physicists of the 20th century, specifically writes that this is not the case, as explained in Chapter 4. Thermal energy is a physical state of matter. It does not exist in little blobs, as matter does. If you subdivide a ball of matter that is at some temperature, all the pieces will have the same temperature to start with, and all the pieces will have the same thermal energy, no matter how small or how large they may be. If you immediately reassemble the ball by putting all the pieces back together again, the temperature of the ball will be the same as it was before, and the thermal energy of the ball as a whole will be the same as the thermal energy of any piece. The energies in the various pieces are not additive.

The reason for this is that thermal energy, like temperature, is an intensive property. As explained in Chapter 4, it is a function of the frequency and amplitude of microscopic thermal oscillations of the bonds that hold the matter together. Thermal energy exists in the form of a field of radiation, similar to a gravitational field, that radiates outward from any object whose temperature is above absolute zero. The intensity, or amplitude, of the field diminishes in inverse proportion to the distance from the radiating object, but its frequency—its chemical energy signal—remains constant regardless of distance. The thermal effect of such an energy field upon a receiving object depends entirely upon that object's position within the field relative to the transmitting object.

Planck's law, plotted in Figure 10.1 (also in Figure 4.6 on page 56), was derived empirically—the result of trial and error—to calculate the intensity (amplitude) of radiation as a function of frequency across a broad spectrum of frequencies that is observed to be emitted by a body of matter at a given temperature. Planck's law shows that when

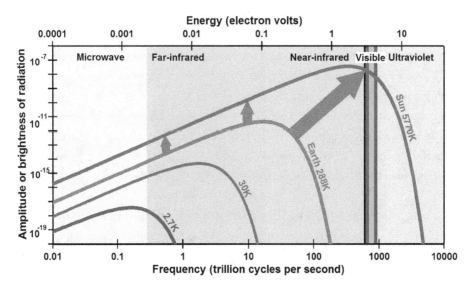

Figure 10.1 When you increase the temperature of a body of matter, you increase the amplitude or brightness of oscillation at every frequency (illustrated by the two small green arrows at two specific frequencies), and you increase the frequency of the maximum amplitude or brightness (large green arrow).

you increase the temperature of a body, you end up increasing the amplitude of the thermal oscillations—currently thought of as the spectral radiance, as discussed in Chapter 4—at each and every frequency (illustrated by two small green arrows at two specific frequencies). You also end up increasing the frequency of the highest amplitude oscillations (large green arrow). Thus, radiation from a given thermal body does not contain high enough amplitudes at each frequency to heat the same body. You can only heat a body radiatively with "hotter" radiation—with radiation from a hotter body.

Heat is well observed to flow from high temperature to low temperature both via **conduction** within matter and via radiation through air or space. At the microscopic level, heat flows in matter via **resonance** from higher amplitudes to lower amplitudes of thermal oscillation at each frequency. A radiation field contains the same distribution of amplitudes of oscillation as the distribution of amplitudes of oscillation on the surface of the radiating matter. As mentioned, the amplitudes of radiation (brightness) decrease as a function of the distance the radiation travels. The "color temperature" of radiation, which is associated with the frequency that has the greatest amplitude, however, will always be equal to the temperature at the surface of the radiating matter as it propagates away. Where radiation that is "cooler" than Earth contacts Earth's surface, it has lower amplitudes of oscillation at each frequency, as well as a lower frequency range, so that heat cannot flow at the microscopic level from the "cooler" radiation to the warmer matter. In other words, the "cooler" radiation is incapable of inducing higher vibrational frequencies in the bonds of the receiving matter than those at which the bonds are already vibrating.

The concept of **radiative forcing** plays a major role in climate science and in all of the Physical Science Basis reports produced by the IPCC since 1990. It is central to the greenhouse warming theory. Radiative forcing is based on the concept "that energy comes *in little blobs of a definite amount*" that can be added together and, as we have just seen, is therefore incorrect. Chemical energy in radiation is equal, at each frequency, to the frequency times the **Planck constant**.

Does the Greenhouse Effect Slow the Rate of Cooling of Earth?

A fundamental assumption concerning the greenhouse effect is that absorption of infrared energy by greenhouse gases increases the temperature of the **troposphere**, decreasing the rate of cooling of Earth. As explained in Chapter 4, however, infrared energy is absorbed by less than 1% of the molecules in the gas as **internal energy** that increases amplitudes of oscillation only at certain specific frequencies of the normal modes of oscillation of the bonds holding the molecules together (Figures 4.8 on page 65 and 10.2). This process has little effect on the temperature of air, however, which is a function of the mean

kinetic energy of all of the molecules making up the air, of which greenhouse gases are very minor constituents. The efficiency of conversion of internal thermal energy within this small component of greenhouse-gas molecules to the external kinetic energy of all molecules in the gas, through numerous collisions, has not been studied in detail. The only experiments that I have been able to find documented in the literature were reported by Knut Ångström[295] in 1900, who showed that any temperature increase resulting from increases in carbon dioxide concentration is minimal.

Whether the temperature increase is small or large, though, is not very important because the primary mode of heat transfer through the troposphere is not by radiation, but by convection. Not only does warm air rise, spawning convection, but weather and ocean currents are driven by the temperature difference between the warmer tropics—heated very efficiently by Sun—and the cooler polar regions—heated much less efficiently by Sun. Far more thermal energy can be moved throughout the troposphere by convection in a storm than can be moved via thermal radiation. The greenhouse effect does not appear to slow the rate of cooling of Earth substantially.

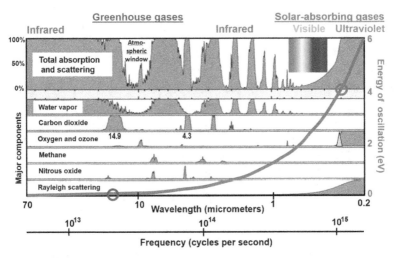

Figure 10.2 The energy of solar ultraviolet radiation reaching Earth when ozone is depleted (red circle, 4 eV at 0.31 μm wavelength, as shown in Figure 5.4, has 48 times the energy absorbed most strongly by carbon dioxide (blue circle, 0.083 eV at 14.9 micrometers (μm) wavelength). Shaded grey areas show the bandwidths of absorption by different greenhouse gases. Current computer models calculate radiative forcing by adding up the area under the broadened spectral lines that make up these bandwidths. Net radiative energy, however, is proportional to frequency only (red line), not to amplitude, bandwidth, or amount.

Does the Greenhouse Effect Violate
the Second Law of Thermodynamics?

The second law of thermodynamics, formally stated in terms of **entropy**, which is hard for most of us to understand, says, in effect, that heat flows spontaneously only from high temperature to low temperature or, in other words, energy cannot be transferred from a body at a lower temperature to a body at a higher temperature. It is well known that greenhouse gases absorb infrared energy along spectral lines (Figure 4.8 on page 65) having very narrow bandwidths (Figure 10.2).

Thus, they are only absorbing a very small portion of Earth's radiation. It is widely assumed that this absorption causes warming of air containing these greenhouse gases. As explained in Chapter 4, however, this has not been proven. It is then widely assumed that these warmer gas layers radiate this energy back to Earth's surface, warming Earth.[296] Gases, however, are not black bodies, which are defined as perfect absorbers and emitters of radiation. Gases only absorb limited frequencies, and what they emit is not well defined, as discussed in Chapter 4.

Figure 10.3 shows a diagram from Trenberth and Fasullo[297] describing the balance of energy in the atmosphere and concluding that 333 W m⁻² of radiation is radiated from

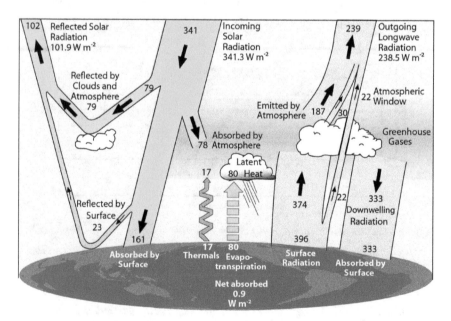

Figure 10.3 The global annual mean energy budget for Earth from 2000 to 2005 in watts per meter squared (W m⁻²) according to Trenberth and Fasullo. As explained in the text, this graph is based on modeling that does not calculate energy correctly.

greenhouse gases back to Earth, while only 161 W m^{-2} of energy comes from Sun. This flies in the face of our common experience that we get hotter standing in sunlight bathed in ultraviolet and **visible radiation** than we do standing outdoors at night, bathed in infrared energy radiated by Earth and thought to be absorbed and re-radiated by greenhouse gases in the atmosphere. Since the atmosphere is colder than Earth, this diagram says that heat is transferred from a colder atmosphere to a warmer Earth. That is equivalent to the expectation that if you walk up to a stove that is colder than you are, you will get warm. We all know that is not correct. The assumption that heat flows by radiation from a colder atmosphere to a warmer Earth violates the second law of thermodynamics and is therefore physically impossible.

Is Sufficient Energy Absorbed by Greenhouse Gases to Cause Global Warming?

Greenhouse warming theory claims that infrared energy absorbed by increasing concentrations of greenhouse gases heats Earth more than increased ultraviolet-B radiation from Sun reaching Earth when ozone is depleted. Radiation subroutines used in all computer models calculate thermal energy to be proportional to the square of the amplitude of light waves integrated across frequency (wavelength), as described in Chapter 4. In this way, climate modelers conclude that there is more energy in the 4.3 and 14.9 micrometer carbon dioxide wavebands labeled in Figure 10.2 than there is in the narrow orange band of ultraviolet at 0.31 micrometers (10^{15} cycles per second) (red circle).

As I explain in Chapter 4, however, chemical energy (E) in matter and in radiation is equal to the Planck constant (h) times the frequency (ν, the Greek letter nu) (red line in Figure 10.2). The increased ultraviolet-B radiation that reaches Earth when ozone is depleted (red circle in Figure 10.2 at 0.31 micrometers) is 48 times more energetic—48 times "hotter"—than the infrared energy absorbed within the broadest bandwidth of absorption by carbon dioxide (blue circle in Figure 10.2 at 14.9 micrometers). There simply is not enough energy involved with Earth's feeble infrared radiation and greenhouse gases to have much effect on temperature compared to Sun's potent ultraviolet energy.

There are many issues with the details of the greenhouse effect that are based on assumptions that do not stand up to scrutiny. Many are discussed in Chapter 4, yet these are largely irrelevant, given this 48-fold difference in energy, the second law of thermodynamics, and the impossibility of radiation from a thermal body raising the temperature of the same thermal body. The greenhouse effect is simply not physically possible.

I have given many examples in this book from throughout geologic history that show how ozone depletion, primarily caused by volcanic eruptions, fits very closely with detailed observations. Let's look at a few examples of how the greenhouse effect does not seem to fit geologic observations very well.

Do Concentrations of Carbon Dioxide Increase Before Temperatures Increase?

During ice ages, carbon dioxide concentrations have increased and decreased at nearly the same time as global surface temperatures (Figure 10.4).[298]

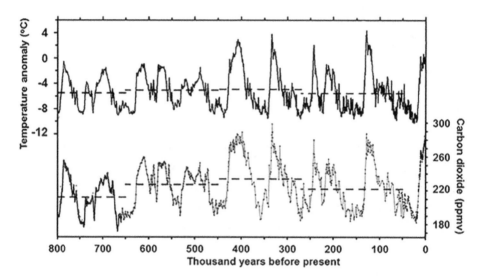

Figure 10.4 Carbon dioxide concentrations and the proxy for temperature measure in Antarctic ice cores track each other quite closely over the past 800,000 years.

Many people see this as confirmation of greenhouse warming theory, yet it is not clear which came first: increases in carbon dioxide causing increases in temperature by the greenhouse effect, or increases in temperature warming the ocean and thereby causing increases in carbon dioxide concentrations in the atmosphere due to the reduced solubility of carbon dioxide in warmer water. Most detailed studies suggest that carbon dioxide levels before the 20th century rose after temperature increased,[299] although a few disagree.[300] The bulk of the data implies that atmospheric concentrations of carbon dioxide may simply be a function of ocean temperature, but the data are not yet definitive because of the difficulties in determining precise relative timing. One thing is clear, however. During the cooling phases of the glacial cycles, reductions in atmospheric carbon dioxide concentration appear to lag behind decreases in temperature by decades to centuries (Figure 10.5).[301]

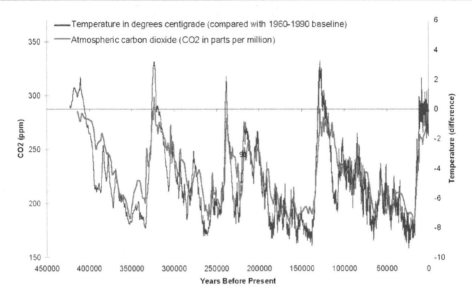

Figure 10.5 During the cooling phase of glacial cycles, atmospheric concentrations of carbon dioxide (red) appear to lag temperatures (blue).

This suggests that a given atmospheric concentration of carbon dioxide cannot prevent a precipitous decrease in global temperature, an observation that, in itself, appears to refute greenhouse warming theory.

Are Increases in Temperature and Carbon Dioxide Concentrations Contemporaneous throughout Geologic Time?

The relationship between rising temperature and rising concentrations of carbon dioxide is not consistent throughout most of geologic time. For example, Figure 10.6 shows the **oxygen isotope** proxy for temperature, the black curve, based on deep-sea **foraminifera** over the past 40 million years (Ma).[302] The blue curve shows the **partial pressure** of atmospheric carbon dioxide (**pCO2**) measured in sediments from cores taken at Ocean Drilling Program Site 925 in the western equatorial Atlantic Ocean.[303] While both curves show a decrease over the past 40 million years, the times and amounts of decrease are quite different. There is little basis here for arguing that pCO_2 causes warming. In detail, note that the substantial warming and/or deglaciation between 27 and 23 million years ago (Ma) (the Oligocene warming) is accompanied by a long-term decrease in pCO_2, the exact opposite of what is predicted by greenhouse warming theory. Also, at 23 Ma, there is major positive excursion (downward, note inverted y-axis) of the oxygen isotope data ($\delta^{18}O$), indicating cooling, coinciding with a very slight up-tick in pCO_2, again defying

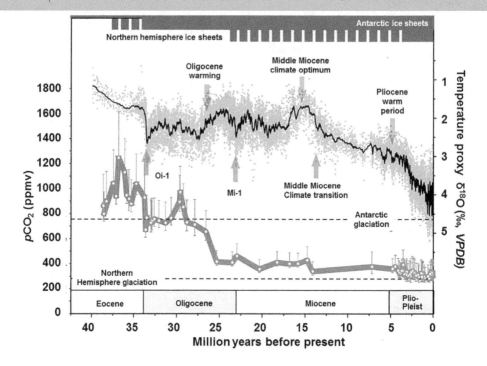

Figure 10.6 The relationship between temperature and carbon dioxide over the past 40 million years is not obvious. Red arrows show times of major warming. Blue arrows show times of major cooling. The red bars show approximate times of Antarctic and Northern Hemisphere ice sheets. The dashed lines are glacial thresholds deduced from climate models.

theory. Zhang writes "Our records suggest invariant carbon dioxide concentrations during this apparent glaciation/deglaciation, defying our current understanding of the necessary forcing required to drive Antarctic ice sheet variability." Note that pCO_2 values were close to current values during the Miocene (23-5 Ma), even though temperatures were substantially higher than they are today.[304] Global temperatures in the late Miocene with pCO_2 less than 350 ppmv exceeded those in the early Pliocene with pCO_2 greater than 350.[305] Furthermore, "why did comparably low carbon dioxide values 22 million years ago not initiate glacial cycles?"[306]

The four major glacial epochs over the past 600 million years did not occur at the same times as low levels of carbon dioxide. The green shaded area in Figure 10.7 shows the oxygen isotope proxy ($\delta^{18}O$) for tropical sea surface temperature.[307] Values below the horizontal green line show times of ice ages; values above the line show warmer times. The blue curve shows sea level.[308] The red curve shows the ratio of the atmospheric concentration of

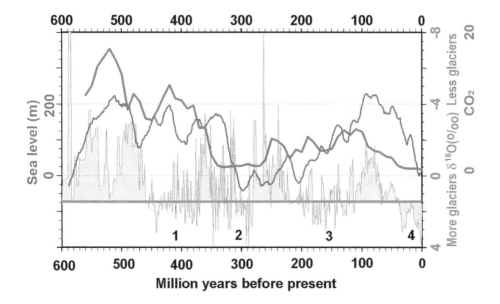

Figure 10.7 Carbon dioxide concentrations were as much as 13 times current concentrations during glacial epoch 1 and as much as 6 times current during glacial epoch 3.

carbon dioxide to current concentrations based on a detailed model for the weathering of volcanic and other rocks[309] with some calibration from pore spacing in fossil leaves.[310] The glacial epochs (numbers 1 to 4) are the only times when geologic evidence for glaciation is widespread over the globe.[311] Note that carbon dioxide concentrations were 9 to 13 times greater than current concentrations during glacial epoch 1, 1.4 times greater for epoch 2, and 3 to 6 times greater for epoch 3.

The correlation of relative carbon dioxide concentration with sea level is somewhat better, which makes sense in that when the oceans are warmer, they do two things: expand, and release carbon dioxide.

Other Problems With the Greenhouse Effect

Concentrations of carbon dioxide in the atmosphere decrease following major **explosive** volcanic eruptions even though carbon dioxide is the second most voluminous gas erupted by these **volcanoes** after water vapor. Pinatubo, in 1991, erupted 42 to 234 megatons of carbon dioxide,[312] but the annual increase in carbon dioxide concentrations per year at Mauna Loa Observatory decreased from 1.4 ppm the year before the eruption to 1.0 the year after the eruption, to 0.7 the second year after the eruption, rising to 1.2

the third year, and to 1.9 the fourth year. The most straightforward explanation is that the well observed cooling of 0.9°F (0.5°C) for a couple of years, caused by the **aerosols** that formed in the lower **stratosphere**, cooled the ocean surface so that it was able to absorb more carbon dioxide.

The sensitivity of climate to a doubling of carbon dioxide is not observed in the laboratory or in Nature; it is calculated on the assumption that most warming during a certain period of time was caused by greenhouse gases.[313] As I have shown, however, the case for greenhouse warming is weak on many levels, whereas ozone depletion plays a substantial role in warming Earth's surface and must be taken into account when pursuing carbon dioxide sensitivity studies.

The red line in Figure 10.8 shows the normalized mean monthly temperature anomalies in the Northern Hemisphere for the period from 1975 to 1998 compared to a base period from 1951 to 1974.[314] The green line shows the normalized, inverted mean monthly anomalies in ozone concentration (ozone depletion) for the same period relative to the same base period. The blue line shows the normalized, mean monthly anomalies in carbon dioxide concentration measured at Mauna Loa, Hawaii, for the period 1961 to the present. The well-defined peak in ozone depletion in March is contemporaneous with the equally well-defined peak in the temperature anomaly, and their trends are well-correlated (r=0.92), suggesting a possible causal relationship. The

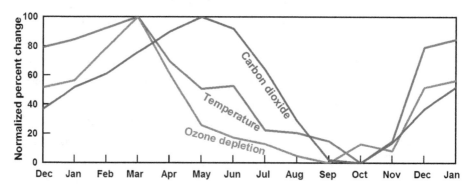

Figure 10.8. Mean monthly values of northern hemisphere temperature anomalies (red) and ozone depletion (green) for the period 1975 to 1998 and of atmospheric carbon dioxide concentrations at Mauna Loa, Hawaii, (blue) since 1961, normalized as percentages. Carbon dioxide anomalies, peaking in May, show only a minor effect on temperature, but coincidence of the peaks in ozone depletion and temperature in March suggest a possible causal relationship. (Graphic and analysis courtesy of David Laing.)

yearly peak in carbon dioxide concentrations, however, occurs in May, two months later than the March peak in the temperature anomaly, precluding a causal relationship, and the blue curve is poorly correlated with the red curve (r=0.54). This simple, observation-based analysis supports the experimental result of Knut Ångström that increases in carbon dioxide do not seem to have much direct effect on increases in temperature and therefore directly contradicts the central assumption of greenhouse warming theory.

Periods of global warming typically start abruptly and are contemporaneous with major increases in **effusive basaltic** volcanism, as discussed in Chapter 8. Volcanic eruptions provide a clear explanation for sudden change, while Rashid et al.[315] conclude "a close examination of paleoclimatic data and modeling results does not show adequate support for many of the widely accepted explanations for abrupt climate change."

Since the concentration of carbon dioxide is essentially homogeneous throughout the globe and increases consistently with time, it is very difficult to explain how changes in carbon dioxide concentrations can cause the clearly observed amplification of warming in polar regions[316] and why the greatest warming occurs during late winter to early spring, when ozone is most depleted.

Hansen et al.[317] show that climate models utilizing carbon dioxide as a primary driver of temperature increase have overestimated global warming since 1998 and are most likely overestimating warming in future decades. Their links to regional climate are also tenuous.[318]

Seidel et al.[319] note that the width of the tropical belt has increased by 2° to 5° in latitude since 1979. Polvani et al.[320] attribute expansion in the Southern Hemisphere to ozone depletion. Allen et al.[321] argue that "recent Northern Hemisphere tropical expansion is driven mainly by black carbon and tropospheric ozone, with greenhouse gases playing a smaller part".

Summary

Between 1975 and 1998, a strong correlation between world mean temperatures and atmospheric concentrations of carbon dioxide (Figure 3.2 on page 33) suggests that increases in carbon dioxide concentration might cause temperature increases, as predicted by greenhouse warming theory. Data from outside of that narrow time window do not seem to fit that model very well, however. As discussed in Chapter 4, absorption by any gas of radiant energy that is insufficient to cause **photodissociation** does not appear to cause much warming. That is confirmed by the observation that the greatest heating in the stratosphere occurs in its upper regions, where photodissociation is dominant, rather than lower down, where the concentrations of ozone are greatest. Observations of climate throughout geologic history and in recent times, do not seem to fit the greenhouse warming theory very well.

CHAPTER 11

WHAT ARE SOME OTHER IMPLICATIONS OF LIGHT BEING A CONTINUUM OF FREQUENCY?

"All these fifty years of conscious brooding have brought me no nearer to the answer to the question 'what are light quanta?' Nowadays every Tom, Dick and Harry thinks he knows it, but he is mistaken."

—A. Einstein, 1951

I imagine that both Max Planck and Albert Einstein, the fathers of modern physics, might have interesting opinions about my primary conclusion in Chapter 4, that light exists in air and space as a field, composed of a spectrum, or more precisely a continuum, of frequencies that originates as the frequencies and amplitudes of asymmetric harmonic oscillations (Figure 4.4 on page 51) of all the degrees of freedom of all the molecular bonds on the surface of a radiating body of matter, and not as waves, particles, or some **wave-particle duality**. With this view, **quantum mechanics** becomes physically more intuitive. Both Planck and Einstein died frustrated that proponents of quantum mechanics made a special point of its not being physically intuitive. How could physics not be physically intuitive? Mathematics can do that, but physics—the science of the physical?

Planck postulated, in 1900, that the chemical energy in radiation (E) was equal to a constant (h) times its frequency (ν, the Greek letter nu) (E=hν) in order to be able to formulate his empirical law describing spectral radiance and color **temperature** (Figure 11.1).

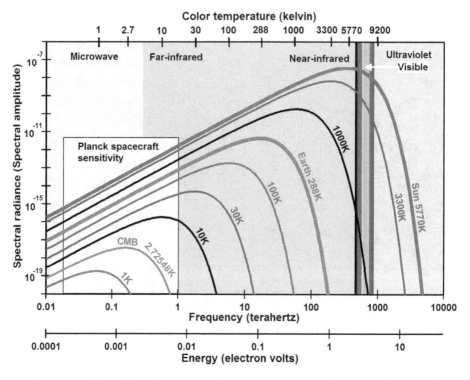

Figure 11.1 Planck's law shows that radiation from a warmer body has higher spectral amplitudes of oscillation at all frequencies than does radiation from a cooler body and exhibits its greatest amplitude at a higher frequency. Each solid line shows the spectral amplitude radiated from a body at thermal equilibrium for the temperature shown. 2.72548K is the temperature of the cosmic microwave background.

He saw this postulate merely as a useful mathematical trick, even though he spent considerable time trying to derive equations concerning radiation in terms of harmonic oscillators. The tradition in physics, since Maxwell in the 1860s, has been that light travels through space as waves, even though the hypothetical **luminiferous aether**, the stuff in space that waves could travel in, had been shown in 1887 not to exist.[322] Nevertheless, the tradition of waves constrained Planck's thinking.

In 1905, Einstein[323] started talking about E=hν as a "light **quantum**." He was trying to understand the photoelectric effect,[324] whereby electrons are emitted when light with sufficiently high frequency interacts with surfaces of certain metals that have been cut or polished to remove oxides. Einstein is likely to have reasoned, although it is not stated in this way in his paper, that if a particle, an electron, is set free, then it was probably knocked loose by another particle. It is this thinking that has dominated particle physics since it was founded—essentially—in 1905. Much modern experimental physics is based around photomultipliers, devices particularly sensitive to the photoelectric effect. Experiments have been conceived, designed, carried out, and interpreted thinking in terms of particles, especially **photons** and electrons, as carefully documented by Brigitte Falkenburg in her book *Particle Metaphysics*.[325] The dirty secret is, however, that we really do not yet understand precisely what a photon is (or if it even exists), or what an electron is or, for that matter, how an atom is actually constructed. We do have a lot of ideas and theories that seem to work fairly well—or at least we think they do.

Energy Equals Frequency Times a Constant

Most physicists should agree that E=hν, commonly called the **Planck-Einstein relation**,[326] correctly describes the energy of an atomic/molecular harmonic oscillator. Note that mass is not included in this equation. Mass plays a role in determining the natural frequencies of the oscillations of the bonds holding matter together, but there is no net change in mass, so that the change in energy is independent of mass. This is entirely consistent with the concept of electromagnetic energy as a frequency field—much like a gravity field—surrounding the emitting object rather than as a "shower of light" transmitting waves or massless particles of energy outward through space and delivering them to a receiving object. Rather, it is simply the receiving object's molecular structure and its position within the frequency field that determines the resonant response of the chemical bonds in the object, whose consequent vibrations generate energy locally, with no inter-object energy transfer required. This concept may not make sense to those traditional physicists used to thinking of light waves, photons, mechanical oscillators with springs, pendulums, balls rolling in curved bowls, and such, but it begins to make a lot of sense when you think of force fields and very high-frequency microscopic oscillations about a minimum in energy between attractive and repulsive electrostatic forces, as shown in Figure 4.4 on page 51.

Now increasing the frequency—and hence the energy—of a bond's oscillation will at some point (E_{max} in Figure 4.4 on page 51) break the bond. There is no collision required. More importantly, this explains why, in the photoelectric effect, no amount of radiation can dislodge an electron until some minimum frequency, and some minimum amplitude, of light is reached. Such thinking about the energy required to break a bond or dislodge an electron helps us understand, quite well, **photoionization**, **photodissociation**, and the structure of Earth's atmosphere, as described in Chapter 5.

The most important point, then, is that frequency of **electromagnetic radiation** in space is both a field and a continuum. It is only quantized when a molecule of gas or solid matter absorbs, via **resonance**, the specific resonant frequencies of the bonds holding the molecule together. Since each frequency has a distinct energy, the molecule extracts from the electromagnetic field only very specific frequencies—energies—determined by the chemical structure of the atom or molecule. All of these different, coexistent energies can be thought of as making up a photon, which is useful as a mathematical shorthand to represent the spectrum of energy induced by the field in the molecule or atom. All the same, these frequencies have no mass, and thinking about them as colliding particles leads to the can of worms known to physics as the Heisenberg uncertainty principle.

We also tend to quantize the continuum of frequency in our thinking when we simply designate some particular frequency. Light is a continuum of frequencies, however, and that is not something we are used to thinking about.

The most distinctive feature of light in air and space is that myriads of frequencies coexist within an electromagnetic field, but they do not interact with each other until matter is present—even just a tiny speck of matter, as is often invoked in the derivation of many equations in quantum electrodynamics. As you look around, every molecule of matter that you can see is oscillating at a specific frequency (color), and those frequencies all coexist in the field that permeates the space between what you see and your eye. We could express this in terms of millions of frequencies, but actually we are talking about a continuum of frequencies. How can you visualize that? That is another distinctive feature of light: you cannot see it until it interacts with matter, such as the rods and cones of your retinas. In the absence of such interaction, light is not visible at all.

This continuum of electromagnetic energy is homogenized as heat in matter by **conduction**. One way or another, the amplitudes and frequencies of ambient radiation are changed via conduction as they interact with matter to form a continuum approximated, at thermal equilibrium, by Planck's law as a function of absolute temperature (Figure 11.1).

Another distinctive feature of light is that the energies are not additive. Since each frequency is an energy divided by the **Planck constant**, and frequencies coexist and are not additive, then the energies associated with light are not additive either. This goes back to Richard Feynman's statement cited in Chapter 4 that "it is important to realize that in physics today, we have no knowledge of what energy is. We do not have a picture that energy comes *in little blobs of a definite amount*."[327] Thinking of energy as a field rather than as little blobs of definite amount will be the most challenging transition to make for most climatologists and physicists as a result of reading this book.

So What Is a Quantum?

A **quantum** in physics is defined in Wikipedia, for example, as "the minimum amount of any physical entity involved in an interaction. Behind this, one finds the fundamental notion that a physical property may be 'quantized,' referred to as 'the hypothesis of quantitization.' This means that the magnitude can take on only certain discrete values." In the photoelectric effect, the effect that led Einstein (1905)[328] to define light quanta, the minimum thing required is energy, which is frequency times a constant ($E=h\nu$). The ambient frequency field must contain some minimum frequency for the effect to happen. Yet frequency, itself, is not quantized, except in labeling when we refer to some specific value. Any specific value or number for frequency in electromagnetic radiation occurs within a broad spectrum of frequencies. Frequency in an electromagnetic field is a continuum. It is not quantized. When we refer to electromagnetic energy as photons or quanta, we are thinking "that energy comes *in little blobs of a definite amount*." Yet Feynman writes specifically that "we do **not** have a picture that energy comes *in little blobs of a definite amount*." What we observe is that frequency is a continuum, but that some minimum frequency, typically in the blue to ultraviolet range, is required to cause the photoelectric effect, the effect observed using photomultipliers that are central to experimental quantum physics.

What Is Quantum Entanglement?

At any point in space, in a gas such as air, in a liquid, or in a translucent solid, a field of light consists of a spectrum (continuum) of frequencies (colors), in which each frequency can be thought of as having spectral amount (amplitude or brightness). Disturbances in this field appear to travel the shortest distance between source and receiver, a straight line that we simplistically think of as a ray, or beam. When we look at colorful fall foliage, for example, each molecule or cell of everything we can see resonates with portions of Sun's radiation field and is warmed much like a pot on a stove. The molecules responsible for color are caused to resonate, expressing a very narrow spectral frequency, which we perceive as a specific color. This radiation has a brightness determined by the brightness of the light illuminating the matter. The brightness goes to zero as the amount of incoming light goes to zero, causing the molecule to cease expressing visible frequencies bright enough to see, although the molecule continues to radiate **infrared** frequencies based on the temperature of the matter that it is part of.

The radiation from each molecule perceived by our eyes, photographic film, or the sensor of a digital camera, for example, causes a photochemical reaction, and the amount of that reaction is defined by the spectral amount (brightness) received at that specific frequency. Our field of view is made up of so many of these "rays" that we can resolve the intricacies of color with very high accuracy, yet these rays of light do not interfere with each other until they interact with matter. We take all this for granted, but it feels like "spooky interaction at a distance"—Einstein's wonderful description of quantum entanglement. The energy of oscillation in one piece of matter over there (a leaf) influences the energy of oscillation in another piece of matter over here (our retinas) at an arbitrarily large distance separated by a gas such as air or by the vacuum of space. This simple "entanglement" is a physical property of common resonance within a pervasive electromagnetic field.

Quantum entanglement, even at speeds much greater than the velocity of light, has taken on far more complicated mathematical properties that are actively debated through the different interpretations prevalent in quantum mechanics, but all of that is in the esoteric playground of the theoreticians. For the rest of us, there is plenty of fairly obvious "spooky action at a distance" going on all around us.

What Is Dark Energy?

Solar energy in space is observed as thermal radiation—an electromagnetic field—the macroscopic manifestation of the very high-frequency oscillations that form an electric field that induces a magnetic field, that induces an electric field, ad infinitum. Radiation travels in space at the speed of light, which must be related to the time it takes these mutual inductions to occur. According to Maxwell's equations, the velocity of light is equal to the square root of $1/\varepsilon_o\mu_o$ where ε_o is the electric constant or vacuum permittivity—the resistance to forming an electric field in a vacuum—and μ_o is the magnetic constant or vacuum permeability—the ability to form a magnetic field in space. Radiation in space does not have a wavelength until it interacts with matter, and even then, wavelength may simply be a mathematical construction that we have traditionally found useful in explaining the reflection and refraction of light. Radiation in space is simply frequency, which, by the Planck-Einstein relation, is a potential to cause heat in receiving matter.

Since radiation cannot be seen until it interacts with matter, radiation in space is a dark continuum of frequencies, that is, a dark continuum of energies. The existence of **dark energy** has been proposed from physical cosmological theory. Since Earth-bound observers can detect less than 5×10^{-8} % of Sun's radiation field, there must be a lot of dark energy in our Solar System. The amount of Sun's dark energy changes over time primarily as the rate of its conversion of mass to energy changes. It also changes in space when this energy interacts with matter, such as casting a shadow behind a planet. Warm black bodies radiate a broad spectrum of frequencies of oscillation that have the potential to heat matter according to Planck's law. Cooler black bodies can absorb this potential, using it to generate heat through resonance, and thus get warmer. Radiation from cooler black bodies appears not to affect[329] warmer black bodies and does not induce resonance in them.

What Is Dark Matter?

Based on observations of the cosmic microwave background by the Planck spacecraft (2009 to 2013) (Figure 11.2)[330] and the standard model of cosmology, the total mass-energy reservoir of the Universe is thought to contain 4.9% ordinary matter, 26.8% **dark matter**, and 68.3% dark energy. Dark matter cannot be seen; it does not appear to resonate with, or emit, electromagnetic radiation but is inferred to exist because of its gravitational effects on ordinary matter. Dark energy is a hypothetical form of energy that theoretically permeates space and is thought to accelerate the expansion of the Universe but is not seen. Many explanations have been proposed for these invisible forms of matter and energy.[331] I stated above that because Earth-bound observers can only detect less than 5×10^{-8} % of Sun's radiation, there must be a lot of dark energy in our Solar System whose percentage changes

Figure 11.2 The Cosmic Microwave Background (CMB) as mapped by the Planck spacecraft observing the Universe at frequencies between 27 gigahertz and 1 terahertz.

in time primarily as the rate of conversion of mass to energy in Sun. From Figure 11.1, it is obvious that when matter cools to temperatures close to absolute zero, it only radiates, and therefore only resonates with, radiation in the far infrared and microwave bands and that much higher frequencies in the near infrared, visible, and ultraviolet wavelengths would pass straight through, making the matter invisible or dark. In other words, dark matter contains no bonds that can resonate with near infrared, visible, ultraviolet radiation, X-rays, or gamma rays. The Planck spacecraft has sensitivity in the microwave and far-infrared (yellow rectangle in Figure 11.1) and therefore is sensitive to radiation from very, very cold matter. What we currently call the Cosmic Microwave Background has a mean thermal black body spectrum at a temperature of 2.72548 ± 0.00057 K.[332] The map of the cosmic microwave background could therefore be the map of dark matter, some of which is colder (blue) than others (red). Doesn't it seem reasonable, then, that the Cosmic Microwave Background, shown in Figure 11.2, is a map of dark matter, in which the colder dark matter is, the older it is?

Is the Universe Expanding?

The red shift of distant light sources has been interpreted to show that the Universe is expanding. Since the Universe is defined as all of time and space and all of its contents, it is not logically clear how the Universe could be expanding. Since matter shrinks as it cools, and our Solar System and our galaxy, the Milky Way, are cooling, and therefore shrinking, is it possible that the Universe is not expanding at all, but that our domain of the Universe is contracting?

What Is Gravity?

Newton's law of gravity states that any two bodies of matter attract each other with a force that is directly proportional to the product of their masses divided by the square of the distance between them. This is structurally the same as Coulomb's law, which states that the electrostatic force between two point charges is directly proportional to the product of the magnitude of their charges, divided by the square of the distance between them. The similarity of these two equations is highly supportive of the concept of electromagnetic radiation as a field, analogous to a gravity field. At close distances, strong electrostatic forces hold the molecules of matter, and matter itself, together. Gravitational attraction is very weak, but the larger the masses, the larger the force. The mass of Sun is 332,946 times the mass of Earth.[333] Therefore, Earth orbits Sun. Isn't it logical that the work done, the energy transferred, to hold Earth in orbit around Sun is analogous to an electrostatic force that is just a little less than E_{max} in Figure 4.4 on page 51? Is gravity simply the electrostatic force that holds together really large masses making up star systems?

Moving On

These are all logical implications of electromagnetic radiation (light) being a field with a spectrum of frequencies, each with an amplitude that diminishes with distance. All of these inferences need to be tested against observations. Are these implications correct? I look forward to the discussion.

I have had the privilege for many years of being friends with one of the top theoretical physicists of the last century. We have talked for many hours, and I am most impressed, watching his mind attack problems. He has helped me grapple with the details of temperature and many other things.

For several years, I tried to convince him that a photon is a very useful mathematical variable, but that it is not a physical thing. He has lived and breathed photons for more than 60 years of his distinguished professional life. He often took me deep into physics to show me how some famous problem or other could not be solved without the photon. He also chided me regularly, saying "you could never pass the exams that all the rest of us physicists have had to pass." Still, he could never counter my fundamental argument that the details of the energy induced by an electromagnetic field in a molecule of gas, something we tend to think of as a photon, is determined by the physical structure of the molecule, at the location of the molecule, and, therefore, did not travel as a photon from Sun or from any other thermal body.

Finally, last fall, in exasperation he said "Peter, if you are right, then everything I have done in my life is wrong." Lesser men might think that, but they would not be able to say it. I was impressed by his honesty and struck by the realization of just how threatening new ideas can be, even though most scientists say they value thoughtful new ideas. I also told him that I do not agree with his conclusion. Many of the greatest minds in physics have collaborated in the development of quantum electrodynamics. They have developed a body of brilliant mathematics that appears to explain observations in some cases to an accuracy of eleven decimal places.

If I am right, understanding this mathematics in the light of a new point of view will lead to new discoveries, new insights, and perhaps an improved standard model and an improved theory of everything. That is the fun of science—let the fun begin.

CHAPTER 12

HOW COULD SCIENCE HAVE BEEN SO FAR OFF THE MARK?

"If at all possible, try it and see what turns up. Theorists almost always dislike this sort of approach."

—Francis Crick, 1988

"Today's scientists have substituted mathematics for experiments, and they wander off through equation after equation, and eventually build a structure which has no relation to reality."

—Nikola Tesla, 1934

The rise and fall of the science of **greenhouse gases** as the primary cause of global warming will most likely be the subject of many papers and books in the future. Here are a few observations that I feel are especially worthy of note.

The Critical Importance of Observations

Science is about making as precise, reliable, and repeatable observations as is possible, measuring and describing the properties of the phenomena observed, and then, based on intuition, logic, experiment, and theory, trying to explain what could have caused those phenomena. The ultimate test is whether predictions made from this understanding and its conclusions can be tested, verified, and replicated. Science without careful observation and experiment is somewhere between fiction and fantasy—it is certainly not science. Theory without careful observation and experiment may spawn some very good mathematics, but it may also be physically irrelevant.

"*Serious experimental work on radiant heat got under way in the decade of the 1770s, and already by 1780 two important results had been obtained. First, radiant heat had been distinguished from 'ordinary heat,' i.e., from convection and **conduction** effects. And, second, radiant heat had been clearly separated from light, although it was known to obey the same optical laws as light.*"[334] Marc-Auguste Pictet—the scientist whose work is discussed in the foregoing quote—observed, in a simple experiment often repeated in classrooms today, that **radiation** from a colder body of matter is reflected rather than being absorbed by a warmer body. Joseph Fourier, in his book *The Analytical Theory of Heat*,[335] published in 1822 and called "a great mathematical poem" by Lord Kelvin,[336] used his first 45 pages to summarize a wide variety of observations on which he would base his mathematics. His book is still widely read today, and these 45 pages are some of the best observations of heat described anywhere in the literature. One special gem is paragraph 40: "*Of all modes of presenting to ourselves the action of heat, that which seems simplest and most conformable to observation, consists in comparing this action to that of light. Molecules separated from one another reciprocally communicate, across empty space, their rays of heat, just as shining bodies transmit their light.*"

In 1859, John Tyndall observed in the laboratory that small amounts of **infrared radiation** are absorbed by "compound gases" consisting of three atoms of two different chemical elements, such as water vapor (H_2O) and **carbon dioxide** (CO_2), but not by gases with only one element, such as nitrogen (N_2), oxygen (O_2), or argon (Ar), which together make up 99.97% of Earth's atmosphere.[337] Tyndall's observations ignited speculation that changes in concentrations of carbon dioxide in the atmosphere might explain the origin and demise of ice ages.

Throughout the last half of the 19th century, many renowned physicists studied the absorption of visible and infrared radiation, discovering the absorption peaks shown in Figure 10.2 on page 143. These were sophisticated experiments, especially those involving the infrared, because of the difficulty of making the observations, and the need, typically, to design new instruments.[338] Many of these old papers still provide interesting reading today. Some of the best work regarding absorption of infrared was done by Samuel Pierpont Langley.

In 1896, Svante Arrhenius, who received the Nobel Prize for Chemistry in 1903 as one of the founders of the science of physical chemistry, hastily wrote a paper based on Langley's data.[339] This paper is often referred to today as the founding treatise on greenhouse warming theory. Looking back in 1997, Elisabeth Crawford wrote: "Arrhenius' final results are impressive both as an innovative exercise in model-building and as a first approximation of the influence of CO_2 on climate. This should not make one forget, however, that they hardly rested on solid empirical ground. Arrhenius did not heed Langley's warning that his investigation had yielded 'no conclusion which we are absolutely sure of.' "[340]

Then, in 1900, Knut Ångström—son of Anders Jonas Ångström, for whom the Ångström unit of wavelength for radiation was named—conducted experimental studies, in the Canary Islands and in the laboratory, of the absorption of infrared radiation by carbon dioxide and by water vapor.[341] Ångström claimed that Arrhenius' use of Langley's data *"can only give rather uncertain results because the difficulties multiply significantly when absorption bands of two different elements, in this case water vapor and carbon dioxide, are one on top of the other and therefore it depends on the separation of the two. ... The Earth's atmosphere would, according to Mr. Arrhenius, even if it is as dry as possible, absorb about 60 percent of earth radiation. The changes in absorption would be colossal with increases in carbon dioxide and a large enough carbon dioxide layer would even fully absorb terrestrial radiation ... Mr. Arrhenius also believes that these findings might be a cause for an ice age. To go into more depths of these findings, which were further developed by Mr. Ekholm [N. Eckholm, Die Zeitschrift "Ymer", 1899], does not seem appropriate as stated above."*

Ångström's experiments were not overly sophisticated, but they were provocative, and they proved sufficient to terminate interest in carbon dioxide as an explanation for ice ages among physicists at the time. Ångström is the only experimental physicist that I know of who actually tried to measure the increase in temperature supposedly caused by an increase in atmospheric carbon dioxide concentration. That is quite a shocking statement. Greenhouse warming theory is founded on the proposition that a geometric increase in atmospheric carbon dioxide concentration causes an arithmetic increase in temperature of the gas, i.e., a doubling (or tripling, etc.) of carbon dioxide will produce so many degrees of warming. This effect should be measureable and quantifiable in the laboratory. True, there might be complications and feedbacks in Nature, but the basic quantification should be clearly observable in the laboratory. Modern physicists have quibbled with Ångström's experiments, but they have not tried to reproduce them or to disprove them experimentally.[342] By any measure, this is a glaring omission in scientific protocol, and one that needs to be addressed.

The greenhouse warming theory was resurrected from the trash heap of history 38 years later, in 1938, by Guy Stewart Callendar,[343] a British steam engineer, and again in 1956 by Gilbert Plass,[344] an American physicist, who was the first to computerize the mathematics for calculating absorption of radiation. Many others contributed. Meanwhile, most experimental physicists became preoccupied with **quantum mechanics** and experimental work related to quantum mechanics. Spectral physicists, however, have done extensive experimental work throughout the 20th century determining the intricate details of absorption of radiation, summarized neatly in the extensive HITRAN database[345] used by climate modelers, but none did experimental work to quantify to what extent air is heated when infrared energy is absorbed by carbon dioxide or other greenhouse gases. It was recognized that the major absorption was by water vapor (Figure 10.2 on page 143) and that water vapor can constitute up to 5% of the lower atmosphere, compared to 300 **parts per million** (0.03% =167 times less) for carbon dioxide, but scientists rationalized this apparent problem away rather than measuring the effects in detail.

Concern over the public policy aspects of greenhouse gases was expressed, in 1965, by a subcommittee of the President's Science Advisory Committee in a report written by Roger Revelle, an oceanographer, Wallace Broecker and Harmon Craig, both distinguished geochemists, Charles Keeling, a geochemist who had established the now classic measurements of carbon dioxide on Mauna Loa in 1958, and Joseph Smagorinsky, an American meteorologist who became the first director of NOAA's Geophysical Fluid Dynamics Laboratory, where the first climate models to study global warming were developed.[346] This report and many others, plus recent experience with drafting the **Montreal Protocol**, led to the development of the Intergovernmental Panel on Climate Change (IPCC) in 1988.

The details would fill several large books, but the bottom lines are that experimental physicists effectively parted company with all the other scientists concerned about climate change in 1900, and that no one in more than 115 years has reported measuring directly how much—if at all—air temperature is increased by a doubling of carbon dioxide concentration. As explained in Chapter 4, carbon dioxide absorbs radiation of a sufficiently high frequency, which generates **internal energy** of oscillations in the bonds holding each carbon dioxide molecule together. Internal energy is distinctly different from the temperature of a gas, however, which is proportional to the average velocity squared, or more precisely to the average **kinetic** energy, of all the molecules making up the gas. The efficiency of the conversion of internal energy within less than 1% of the gas molecules to kinetic energy of all the gas molecules via numerous collisions has still not been studied. Furthermore, the spectrum of radiation absorbed by carbon dioxide in the narrow bands shown in Figure 10.2 on page 143 is quite limited compared to the spectrum of radiation used to define temperature in matter via **Planck's law** in Figure 4.3 on page 48. There are many good reasons to suspect that Ångström's observations were correct, yet no one has ever tried to replicate his experiments as greenhouse warming theory has risen to dominate climate science and related public policy.

Fundamental Questions in Physics

The need for experimental validation is especially important because, as discussed in Chapter 4, the answers to a number of critical questions in physics have been changing over the past two centuries:

- What is heat?
- What is temperature in a gas and in a solid?
- What is energy?
- What is thermal energy?
- How is thermal energy stored in matter?
- Is thermal energy radiated, and if so, how?
- What is radiation?
- How does radiation propagate through air and space?
- Is a gas molecule's ability to absorb radiation limited; i.e., does it get saturated?

These are all very fundamental questions in physics. It took me many years of pondering them in order to reach the understanding described in this book. More critical thought about these basic issues over the past decades would have helped all scientists better understand the actual thermal effects of greenhouse gases.

This need for fundamental thinking was particularly strong because many of the key assumptions regarding climate change were made long before living climatologists were born and long before many important new advances in science were made. For example, in 1862, James Clerk Maxwell[347] wrote down four equations that would soon form the foundation of classical electrodynamics, optics, electric circuits, and, with the addition of quantum concepts, quantum electrodynamics. While these equations are applied to radiation (electromagnetic fields), they are formulated in terms of bond oscillations in matter. The energy in electromagnetic fields induces vibrational energy in matter, which subsequently redistributes it internally and re-radiates it in the form of secondary fields. None of this is obvious, however, because all human interactions with fields are through the matter in our sensors and our transmitters. Consequently, thinking of fields as being similar to waves in matter has been very natural and very useful. All the same, it was also discussed throughout the latter half of the 19th century that there is no matter in space. This led to the search for the **luminiferous aether**, the postulated matter through which light must travel. There were numerous experiments that proved no such aether exists. The issue of what radiation travels in was never resolved, it just took on a new face with the rise of quantum physics, but no one saw fit to ask one basic question: "Does radiation—whatever it is—actually travel at all?" Viewing it as a field, similar to gravity, removes the issue entirely, along with the need for quantum physics!

Another example is from Samuel Pierpont Langley,[348] who did the primary work in the late 19th century detecting and measuring absorption of infrared radiation. He wrote: "the vast amount of the energy in [the infrared] region … is over 100 times that in the ultra-violet." Oh, really? Climatologists, to this day, dismiss the ultraviolet as not having enough energy to be important. Nevertheless, observations detailed in this book show that because the energy in radiation is actually proportional to frequency, and not to bandwidth or wave amplitude, and that the energy in the infrared absorbed most strongly by greenhouse gases is over 48 times less (Figure 10.2 on page 143) than the energy potential in the increased ultraviolet reaching Earth when ozone is depleted. Increases in ultraviolet-B radiation caused by ozone depletion explain most observations of climate change quite directly, clearly, and much better than absorption of infrared radiation by greenhouse gases.

The Role of the IPCC

According to an IPCC press release:[349] *"The Intergovernmental Panel on Climate Change (IPCC) is the world body for assessing the science related to climate change. The IPCC was set up in 1988 by the World Meteorological Organization (WMO) and United Nations Environment Programme (UNEP), endorsed by the United Nations General Assembly, to provide policymakers with regular assessments of the scientific basis of climate change, its impacts and future risks, and options for adaptation and mitigation."*

"The IPCC assesses the thousands of scientific papers published each year to inform policymakers about what we know and don't know about the risks related to climate change. The IPCC identifies where there is agreement in the scientific community, where there are differences of opinion, and where further research is needed."

"The IPCC offers policymakers a snapshot of what the scientific community understands about climate change. IPCC reports are policy-relevant without being policy-prescriptive. They do not promote particular views or actions. The IPCC evaluates options for policymakers, but it does not tell governments what to do.

"IPCC reports draw on the wisdom and dedication of the entire scientific community dealing with climate change, with the involvement of experts from all regions and diverse scientific backgrounds. IPCC authors and reviewers, including the Chair and other elected officials, work as volunteers."

Since 1988, thousands of scientists have volunteered substantial amounts of time—measured in days to months per year—attending IPCC meetings, completing writing projects, and reviewing assignments. Most were already convinced, or became convinced, that the world was warming, that human beings were likely at least partially to blame, that greenhouse gases were most likely the cause, and that if greenhouse gas concentrations were not reduced soon, life on Earth could be in danger. A substantial portion of the professional scientific effort of climate scientists was diverted away from research questioning scientific details and focused via groupthink on building consensus. While consensus is the stuff of politics—of getting people to band together to take action—debate is the stuff of science, getting to the bottom of difficult issues. Under the IPCC's regime, many good scientists morphed into politicians.

Why did the IPCC not point out the need for experimental verification of the assumption that increased concentrations of carbon dioxide actually warm air? Simply put, it was because of human nature. In life and in science, ideas become popular and become strongly defended by groups of people who adhere to those ideas. We are right, and they are wrong—end of discussion. This is the unfortunate foundation upon which many consensus-based groups make decisions. In the absence of a dictator with absolute authority, the majority rules. The safest route through life is to agree with the majority. Throughout most of human history, this has amounted to a matter of life or death. In democratic governments of the 21st century, there is perhaps a bit more room for diversity of opinion, although media-based social bias has increasingly presented truth-seekers with serious challenges.

In science, as in other endeavors, there is a natural tendency to seek group agreement that our ideas are correct. Scientific theories are developed and tested in search of fundamental laws of Nature. Most scientists consciously try, via the scientific method, to be objective, to be rigorous, and to seek truth, yet as Admiral Rickover famously said, "the devil is in the details." Most major scientific issues involve experts from a wide variety of fields. It is practically impossible for any one scientist to be an expert in all relevant fields. We all depend on insights formed by many others, relying on scientific papers vetted through peer-review, on meetings, and on personal interaction with others. We rely on "common knowledge" to help us sort out what is most likely true, what is most important, and what to include in our own world views and in our own research.

All humans not only seek approval, respect, and acceptance from their peers, but also seek unconditional love from their families—their personal tribe. Some need approval just to function. For others, approval enhances the feeling of self-worth. When we are faced with disapproval, we can learn from it, or we can skillfully rationalize it away.

As the former manager of a large segment of a national research program, I know that science, more often than not, depends on group consensus in order to obtain money to fund a program. The program manager depends on group consensus to identify and enable the best qualified researchers to achieve the goals of the program, to review their proposals and evaluate their record of productivity, and to make the most productive decisions.

As Max Planck said in 1936, however, "New scientific ideas never spring from a communal body, however organized." Groupthink stifles new ideas. Groupthink conserves the ideas that its members rely on, and so does peer review. This is only human. Could a communal body be so organized as to question all its assumptions systematically? I would have to agree with Planck that the odds are low, because it takes time for hints of problems to simmer, and it takes freedom and quiet to be able to listen to your intuition. We each travel these roads on different schedules, at different rates, and in different ways. If I had had to sit through committee meetings listening to thoughtful ideas from very different roads, and from very different domains, I might have learned some valuable new perspectives, but my progress discussed in this book would most likely have been delayed. Perhaps most importantly, however, although I need to feel that I am right in order to be really passionate about my work, I also need to be humble and open to the possibility that I might not be right. It is a very delicate balance.

I think Planck describes it well when he says new scientific ideas spring "from the head of an individually inspired researcher who struggles with his problems in lonely thought and unites all his thought on one single point which is his whole world for the moment." My moment has lasted 8 years, so far, and it is still not over.

Consensus Science

Michael Crichton, who received his MD from Harvard Medical School but is more famous for science fiction, said on January 17, 2003, when he gave the Michelin Lecture at The California Institute of Technology:[350] "I want to pause here and talk about this notion of consensus, and the rise of what has been called consensus science. I regard consensus science as an extremely pernicious development that ought to be stopped cold in its tracks. Historically, the claim of consensus has been the first refuge of scoundrels; it is a way to avoid debate by claiming that the matter is already settled. Whenever you hear the consensus of scientists agrees on something or other, reach for your wallet, because you're being had.

"Let's be clear: the work of science has nothing whatever to do with consensus. Consensus is the business of politics. Science, on the contrary, requires only one investigator who happens to be right, which means that he or she has results that are verifiable by reference to the real world.

"In science consensus is irrelevant. What is relevant is reproducible results. The greatest scientists in history are great precisely because they broke with the consensus. There is no such thing as consensus science. If it's consensus, it isn't science. If it's science, it isn't consensus. Period."

The Importance of Critical Questioning and Replication

There have been many good scientists who have disagreed with certain conclusions blessed by the IPCC—with certain details of the science. Some have been able to have a positive effect on the IPCC reports, but they have typically not been encouraged. Some have spoken out in the public arena, and in a few cases have caused change in the majority view. Others have published their findings and moved on. There were many who wondered but decided not to commit the resources, or couldn't find the resources, to pursue their questions. Science depends on a diversity of opinions, on constructive disagreement, and on debate. Unfortunately, it also depends on funding. Many program managers understand the value of variant opinions and have been successful at funding some controversial studies, but energies don't naturally flow in that direction.

Having managed a group of 40 PhD scientists plus 100 support staff at the age of 31, I have had a life-long interest in how to manage creativity—a cat-herding task at best. I concluded years ago that creativity flourishes in the cracks, in the locations where management is limited, in places where the individually inspired researcher has room to grow if he chooses to take charge. Management is often deemed good by how many cracks it seals up. You cannot actually manage creativity, but you can easily snuff it out.

One of the better books about the climate wars that have played out for more than 20 years so far is *The Hockey Stick Illusion: Climategate and the Corruption of Science* by Andrew Montford.[351] There are other opinions on this topic.[352] This was not the most shining hour in science, but it illustrates how well-meaning researchers can get into trouble. I think there are some lessons here for how science could have been so far off the mark.

The hockey stick graph was published by the IPCC[353] in 2001 and is shown in Figure 12.1.

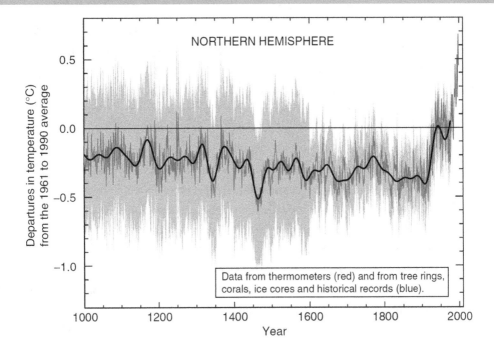

Figure 12.1 The highly contentious hockey stick graph, showing temperature data from thermometers in red and temperature data inferred from tree rings, corals, ice cores, and historical records in blue. Gray shows the 95% confidence range of the blue data. The black line is the running 50-year average of the blue data.

Temperature data from thermometers, plotted in red, have been added to temperature proxies inferred from tree rings, corals, ice cores, and historical records in blue. This graph was recognized immediately as an iconic image demonstrating more clearly than any other, that temperature rise since 1900 has outpaced anything known in the past 1000 years. In other words, warming caused presumably by humans has been much more rapid that anything caused by Nature since 1000 AD.

The fact that many people liked the iconic nature of this plot meant that it and its author were catapulted to fame perhaps before the data could be adequately analyzed. I observed a similar situation in the 1970s when data that were not really clear suggested that an earthquake could occur soon and could injure or kill many people. The popularity of an idea or deep concern about an outcome can, unfortunately, lead scientists to forgo adequate development of the science involved.

In the case of the hockey stick graph, the problem was that in the previous IPCC report, a qualitative plot of temperature over this time period clearly showed the Medieval

Warm Period from around 950 to 1250 AD followed by the Little Ice Age from 1300 to 1850 AD, both of which were well observed historically before thermometers had been invented. One can see these anomalies in Figure 12.1, but the deviation is only 0.1 to 0.2°C, much less than imagined from historic accounts. Some people interpret the historic records to suggest that temperatures were greater during the Medieval Warm Period than from 1900 to 1970. Could the **proxy** data and/or the software used to integrate them into the graph be flawed in some way?

At that time, the battle between people who were convinced that climate was warming and those who questioned that conclusion for any reason was heating up into full-blown warfare. Many wondered about the data and one very persistent person requested access to the data and the computer programs so that he could try to replicate the results. Access was denied by the researcher.

In the scientific method, one of the most important expectations is that novel findings need to be replicated by other independent scientists or laboratories before they are considered valid. Denying access to important data is not an acceptable option in the scientific method. In the past few years, many journals have developed new policies that require the data analyzed for the paper to be stored in publicly accessible data bases before the paper will be published. Such a policy, had it been in effect 15 years ago, might have cooled the climate wars substantially.

In science, critical questioning of all conclusions, critical evaluation of all assumptions, and replication of results, need to be valued and promoted by all.

CHAPTER 13

WHERE DO WE GO FROM HERE?

"The only true voyage of discovery, the only fountain of Eternal Youth, would be not to visit strange lands but to possess other eyes, to behold the universe through the eyes of another, of a hundred others."

—**Marcel Proust**, 1923

We caused the world to warm approximately 1°F (0.6°C) from 1965 to 1998, but it was unintentional. We had no idea that **chlorofluorocarbon gases (CFCs)**, which became very popular in the 1960s for use as spray-can propellants, refrigerants, solvents, and foam blowing agents, could have any effect on climate. CFCs were popular precisely because they were so chemically inert—they did not react chemically with other substances under normal conditions. Scientific research concerning the lower **stratosphere**, however, in part spawned by plans to build a supersonic transport, made it possible for Mario Molina and F. Sherwood Roland[354] to discover, in 1974, that CFCs in the vicinity of unusually cold, **polar stratospheric clouds** can be broken down by **ultraviolet radiation**, as Sun returns to polar regions in late winter and early spring, releasing chlorine atoms that are unusually effective at destroying

173

(depleting) ozone. Molina, Rowland, and Paul Crutzen received the 1995 Nobel Prize in Chemistry "for their work in atmospheric chemistry, particularly concerning the formation and decomposition of ozone" and for their efforts to make the scientific results of their discovery well known and understood.

When the **Antarctic ozone hole** was first observed in 1986,[355] dramatically demonstrating the reality and scope of ozone depletion, scientists and politicians began grasping for an immediate solution. The science was well-enough known to draft the United Nations **Montreal Protocol** on Substances That Deplete the Ozone Layer by 1987 over the protests of the chemical manufacturers. This protocol began to limit CFC emissions upon its initial ratification on January 1, 1989. If scientists and politicians had not worked so effectively and efficiently together, the world's surface temperatures would still be warming—the damage would still be getting worse. Kofi Annan, Secretary-General of the United Nations from 1997 to 2006, said in 2003, that "perhaps the single most successful international agreement to date has been the **Montreal Protocol**."

We are still not out of the woods, however. CFCs are so stable that it will be four or five decades before their concentrations in the atmosphere are likely to have declined to pre-1965 levels. Meanwhile, CFCs continue to deplete ozone, allowing more solar heat to reach Earth, warming oceans, melting glaciers, and increasing the risk of sunburn and skin cancer. Some CFCs are still available on the black market. CFCs still exist in some old refrigerators, air conditioners, and such. We need to eliminate all emissions of CFCs and of all other substances that deplete the **ozone layer**, without exception. We might do well to look for ways of removing CFCs from the atmosphere in polar regions.

The longer ozone remains depleted, the warmer the oceans will become. The oceans are the true thermostat of the climate system. They hold nearly all of the heat content of the climate system. We have moved the thermostat up, and the only natural way to cool the oceans is with more frequent eruptions from **explosive volcanoes**. There is a very active group of scientists who are thinking about geoengineering, including ways to reduce solar radiation reaching Earth. The magnitude of the problem, however, may be far beyond our financial and technical means, and the politics of intentionally changing climate could make the greenhouse warming debate look trivial.

Although we have warmed the world on average close to 1°F (0.6°C) since 1965, and although the world will remain warmer than it was before, we can take comfort in the realization that the major warming predicted by climate models based on greenhouse warming theory has not occurred since 1998 and is unlikely to occur in the future. We can burn all available fossil fuels that we need, as long as we minimize **pollution**, protect the environment, and recognize that the supplies of fossil fuels are limited. We can be truly relieved that predicted climate disaster has been averted through a new understanding of what really causes global warming. Nevertheless, although the predicted continued warming throughout the 21st century is not likely to happen, climate disaster should remain an ever-present concern, especially as world population grows, straining available resources.

World Population and the Escalating Need for Resources

More than 7.3 billion people inhabit Earth, and that number is predicted to reach 9 billion within 25 years (2040). World population was only 1 billion in 1804, 3 billion in 1959, 5 billion in 1987, and 7 billion in 2011. Such rapid growth strains available resources for food, water, clean air, energy, and many critical minerals required for modern technologies. Furthermore, all people seek improved quality of life, further straining resources because quality of life depends on large amounts of affordable energy. The average American uses 34 times more energy than the average citizen of Bangladesh, 13 times more than a citizen of India, and 4 times more than a citizen of China.[356] We need all available sources of energy just to meet growing demand. This is why cutting back greenhouse-gas emissions enough to reduce predicted greenhouse warming has never really looked practical. People will not vote for politicians who will make them reduce their quality of life unless they are absolutely convinced that there is no choice and that limits in energy are applied fairly across the board. Failure to understand this dynamic has led to the collapse of many past civilizations.

We do need to be realistic about the reality that fossil fuels are limited and that alternative sources of energy will become more and more important over the next few decades. Calculating current rates of fossil fuel use against current known reserves suggests that we will run out of oil around 2067, natural gas around 2069, and coal around 2127.[357] The US has 27% of all known coal reserves, while Russia has 18% and China 13%. New technologies keep extending these dates, but fundamental changes in fuel sources will be needed within our children's and grandchildren's lifetimes. Meanwhile, conservation of energy will help substantially. We currently reject (waste) 60% of energy produced.[358] A certain amount of waste is inevitable, but we can improve our energy efficiency significantly.

We also need to burn energy much more cleanly. While most developed countries have reduced air pollution and associated acid rain very effectively since the early 1970s, the World Health Organization estimates that "in 2012, around 7 million people died—one in eight (12.5%) of total global deaths—as a result of air pollution exposure."[359] Air pollution is particularly bad in rapidly industrializing China, India, and much of Asia, home to more than 60% of the world's population. Some days, the pollution reaching southern California from Asia and from ships at sea, is greater than the total allowed to be produced within California by the Environmental Protection Agency.[360]

All countries have, or can acquire, affordable technology to burn fossil fuels cleanly, to minimize pollution, and to meet escalating energy needs in the short term. We can improve this technology. If, for economic reasons, certain countries choose not to use this technology, then the quality of life of their citizens, and perhaps that of others in surrounding countries, will be diminished, and millions will continue to suffer illness and early death because of excessive pollution.

Increased Need for Understanding

While it is clear that the **halogen** gases, chlorine and bromine, emitted by **volcanoes**, cause ozone depletion, we do not yet understand all of the precise chemical paths involved. We also do not have a precise idea about the magnitude of the effects of these gases across the range from quietly degassing volcanoes to small eruptions to large eruptions, or about the regional versus global effects as a function of the latitude of the erupting volcano.

Perhaps one of the most important needs for research involves integrating our rapidly increasing understanding of the magnitude and distribution of ozone depletion as a function of latitude and season into forecasting regional weather. This could have very direct effects on our quality of life and on devising ways to adapt.

The most important need, however, is to try, with a sense of humility, to understand each other and to appreciate what happened to the link between science and public policy over the past few decades. Science provides the most logical, rational way to inform public policy with the most reliable observations and interpretations of Nature. Science should not set public policy, but in today's increasingly technological world, it would likely be suicide not to utilize science to inform decisions regarding public policy. Humans have endured millennia of public policy without science. This has led to some interesting history, but history entailing much loss of life and suffering. Today, however, our prosperity and our very survival depend increasingly on rapidly advancing technology. We need the best information available in order to make increasingly complex decisions.

The "Climate Wars" have gotten down, in the extremes, to a battle between "the scientists," who have become convinced that they understand climate change extremely well, and "the skeptics," who range from scientists who disagree with the majority view to nonspecialists unconvinced that the world is getting warmer, unconvinced that man could be causing climate change, or concerned that doing anything about greenhouse-gases would bankrupt them, their businesses, their nations, or the world economy. Many skeptics, and even some political parties, have been more than happy to renounce science when it doesn't agree with their beliefs—to make public policy based on opinion rather than on our best understanding of Nature. The problem is that the scientists have taken a detour. Many scientists have been claiming that the science of greenhouse warming is "settled," and that we need to get on with reducing **greenhouse gases** before it is too late. Science, however, is never "settled." That very notion is the antithesis of good science. Scientists seek perfection, but the road to perfection twists and turns and sometimes gets detoured. The conservative view of climate change has delayed action long enough for the science to self-correct. If it had not, we would already be wasting major amounts of money reducing greenhouse gas emissions. This is where humility comes in. As we travel through life together, we need to recognize that sometimes our "enemies" are right for the wrong reasons. Sometimes we are not as right as we believe we are. Sometimes it is not worth going to war over things we believe in fervently. Sometimes we just have to admit that we are humans stumbling forward together as best we can, but the important thing is that we move forward.

We can expect that climate will change as rapidly on long-term time scales as weather does on short-term time scales. Predicting weather and climate has great practical and economic relevance to our lives. Evolution of humanity and of all creatures on Earth has been driven by climate. There have been long periods of little change and short periods of rapid change. We need the scientists to help us understand the risks and what we can do about them.

Climate Disaster Is Never Far Away

The most fundamental conclusion in this book is that throughout the long history of Earth, volcanoes have ruled climate. Climate change has been determined primarily by the balance between the number of large explosive volcanic eruptions per century and the volume and duration of effusive volcanism.

Global cooling has been caused primarily by major explosive volcanic eruptions that eject megatons of water vapor and **sulfur dioxide** into the lower stratosphere, where they form aerosols that reflect and scatter sunlight, cooling Earth. Currently, the largest of these types of eruptions happen two or three times per century (Pinatubo in 1991, Katmai/ Novarupta in 1912, and Santa Maria in 1902). When they become more frequent, they will cool Earth into a new ice age.

Global warming, on the other hand, is caused by ozone depletion, which is caused primarily by **effusive, basaltic** eruptions that extrude black, fluid, basaltic lava over large areas. The amount of warming is determined by the volume of lava erupted. The effect of this warming on ocean temperatures, the primary reservoir of heat at Earth's surface, increases with the duration and continuity of the eruption. We truly live at the mercy of Nature.

As David Keys explains in his book Catastrophe,[361] "in AD 535/536, mankind was hit by one of the greatest natural disasters ever to occur.... It blotted out much of the light and heat from Sun for 18 months and resulted, directly or indirectly in climatic chaos, famine, migration, war, and massive political change on virtually every continent." This event was the most severe and protracted cooling event in the past 2000 years (Figure 13.1)[362] and has been attributed to Rabaul volcano in New Guinea, Krakatau volcano in Indonesia, and even to a meteorite impact. The best data, however, suggest that it was an eruption from the 28 mi² (72 km²) Ilopango **caldera**, just outside of San Salvador, El Salvador.[363] It caused world-wide famine, the beginning of the Dark Ages, and the decline of Teotihuacán, a pre-Columbian Mesoamerican city located today 30 mi (48 km) northeast of Mexico City. This was the largest explosive eruption in written history.

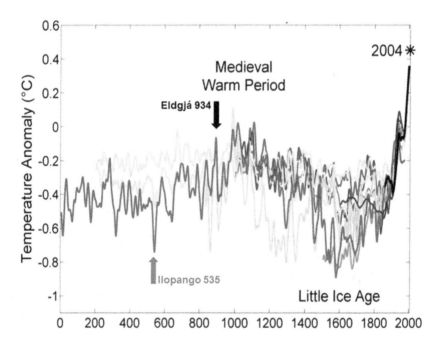

Figure 13.1 Northern hemisphere temperature reconstructions from 0 to 2000 AD. Red arrow shows the 535 AD eruption most likely from Ilopango crater in El Salvador. Black arrow shows the 934 AD eruption of Eldgjá in Iceland. Instrument temperatures in black. More recent reconstructions are plotted towards the front and in redder colors, older reconstructions appear towards the back and in bluer colors.

In 934 AD, Eldgjá erupted in Iceland, the largest effusive flood basalt eruption in written history, extruding 4.3 mi³ (18 km³) of lava over an area of 309 mi² (800 km²),[364] and warming the world into the Medieval warm period (Figure 13.1). When Eyjafjallajökull erupted in Iceland in 2010, airspace was disrupted in northwest Europe from April 15 to 21 and in May, even though only 0.06 mi³ (0.25 km²) of **tephra** was erupted.[365] It is sobering to imagine how horrific the effects of an eruption more than 100 times larger would be today.

Toba volcano in Indonesia, exploded 73,000 years ago (plus or minus 4000 years) belching 672 mi³ (2800 km³) of **magma**,[366] and was possibly the cause of a bottleneck in human evolution, in which as few as 10,000 people were left alive.[367] The Yellowstone Caldera, in northwestern Wyoming, exploded 590 mi³ (2450 km³) of magma 2.1 million years ago.[368] The Siberian Traps in Russia, an effusive eruption, extruded up to 1,000,000 mi³ (4,000,000 km³) of basalt 250 million years ago. Imagine the effects of any of these eruptions on an overpopulated world. Eruptions of this size luckily are not frequent,[369] but they will occur.

There will also be many more eruptions that will be not quite as big, but big enough to be very damaging. While volcanologists may be able to give us some warning, these climate events will change the course of history as they have in the past. Are we prepared?

Listening to Earth

Is there a take-away message in all this? For me, it continues to be a conviction that I have held for the fifty-four years during which I have been enthralled with Earth science. Whenever I feel the need to know and understand something about Earth, I go straight to the source. I may well, and doubtless will, consult books, journal articles, and fellow scientists for opinions on the matter, but my bottom line has always been that this incredibly wonderful planet Earth, on which we are all so privileged to live, is itself our best teacher. There simply is no better substitute for hard evidence based on careful observation. Conceptual models of how Earth works are all well and good, but no matter how clever and sophisticated they may be, they must always bow to on-the-ground truth. I applaud the impressive works of great theoreticians, but in the end, I will always turn to Earth for the final word.

"Speak to the Earth and it shall teach thee."[370]

Figure 13.2 *The portal over the door of Schermerhorn Hall at Columbia University, home for the Geology Department for many, many decades.*

GLOSSARY

Definitions of technical words as used in this book. First appearances of these terms in each chapter are printed in **bold type**. Terms that are cross-referenced within this Glossary are printed in *italics*. Sources include Wikipedia, Wiktionary, and Laing, D., *The Earth System: An Introduction to Earth Science*, 1991, Wm C Brown, 590 pp. Others are original with the author or editor. All definitions have been written or modified to conform with their uses in this book.

Aerosol = Fine solid or liquid particles dispersed and suspended in air.

Albedo = The ratio of the intensity of light energy that is not absorbed (i.e., is scattered or reflected) by a surface to the intensity of light that falls upon it.

Amplitude of oscillation = The maximum displacement in meters during one cycle of one oscillation of one *degree of freedom* of one of the bonds that hold matter together. The *amplitude* increases with the *temperature* of the matter. Amplitudes and frequencies of bond oscillations on the surface of matter induce *electromagnetic radiation* that contains the same amplitudes, but these decrease inversely with the distance traveled as the radiation spreads out from the surface. Amplitude corresponds to the intensity of light (radiation). *Frequency of oscillation* corresponds to the color of the oscillation and does not change with distance traveled in air or space, except for *Doppler effects*.

Andesite = A grayish-green to brown, fine-grained, *extrusive* volcanic rock, more evolved (i.e., has assimilated more crustal materials) than *basalt* and less evolved than *dacite*, and consisting mostly of intermediate plagioclase feldspar (sodium calcium aluminum silicate) with lesser pyroxene and/or hornblende.

Anoxic interval (Anoxic event) = A time period in which large portions of Earth's ocean basins became depleted of oxygen.

Antarctic ozone hole = Every Antarctic spring (September to November), the amount of ozone measured since 1986 has been depleted by as much as 60% over an area from 5.4 to 10.4 million square miles (14 to 27 million km²) covering Antarctica and much of the surrounding ocean. The annual extent of the ozone hole is shown at earthobservatory.nasa.gov/Features/WorldOfChange/ozone.php.

Anthropogenic = Caused by humans.

Ash = Finely pulverized rock and/or lava, consisting of particles less than 2 mm in diameter, produced in a volcanic eruption.

Axial tilt = The inclination in degrees of Earth's rotational axis (23.4 ±1.25°) from a perpendicular to its orbital plane.

Basalt = A primitive, *mantle*-derived, dark gray to black, fine-grained, *extrusive* volcanic rock consisting mostly of calcium-rich plagioclase feldspar (calcium sodium aluminum silicate) with lesser pyroxene and/or olivine.

Birefringence = A property of an anisotropic (properties vary with direction), transparent, crystalline material whereby the refractive index depends on the polarization direction of light.

Bølling warming = An interval from about 14,600 to about 14,100 years ago, in which annual *temperature* over Greenland suddenly rose by as much as 21°C before slowly returning to ice-age levels.

Caldera = A large depression caused by volcanic explosion, by collapse of the crust into a partially evacuated *magma chamber*, by erosion, or (as is usually the case) by various combinations of the three.

Carbon dioxide = A linear, nonpolar, triatomic molecule consisting of a single carbon atom covalently bonded (i.e., sharing electrons) between two oxygen atoms. Carbon dioxide is a gas at standard temperature and pressure and it is regarded as a *greenhouse gas* that absorbs *infrared radiation* from Earth, but absorption only occurs at the specific *resonant frequencies* of the bonds that hold the molecule together.

Catalytic reaction = A chemical reaction whose rate is increased by the presence of a chemical (a catalyst) that does not undergo any permanent chemical change.

Chlorofluorocarbon (CFC) = Any of a class of man-made, gaseous chemicals derived from the aliphatic hydrocarbons methane, ethane, and propane, in which normal hydrogen atoms have been replaced by chlorine and fluorine atoms.

Conduction = The intermolecular transfer of heat energy, in which increased thermal motions are passed from hotter molecules to adjacent, cooler ones.

Contiguous layers = Layers that touch one another. The number of contiguous layers of ice that contain volcanic *sulfate* is a measure of how continuous volcanism is, and it serves, therefore, as a measure of the effect of sulfate on warming or cooling the oceans.

$\delta_{18}O$ (delta O-18) = The ratio in parts per thousand of the heavy oxygen-18 isotope to the normal oxygen-16 isotope compared to a standard. Commonly used as a measure of the *temperature* of precipitation. Used as a *proxy* for temperature in ice cores, *foraminifera*, and corals dating from thousands to millions of years ago. The ratio is lower during glacial ages and higher during warm episodes.

Dacite = An *extrusive* volcanic rock, lighter in color and density, more evolved (has assimilated more crustal materials) than *andesite*, and richer in quartz and high-sodium plagioclase feldspar (sodium calcium aluminum silicate).

Dansgaard-Oeschger event = One of 25 sudden, but temporary $\delta^{18}O$ *proxy temperature* increases of up to 24°C in Greenland ice core records, occurring between 120,000 and 10,000 years ago at random intervals that average 1470 years apart. Warming out of ice-age conditions normally occurs in less than 3 decades, while cooling back into ice-age conditions typically takes many centuries.

Dark energy = Energy that is not visible but is hypothesized to permeate all of space and to accelerate the apparent expansion of the Universe.

Dark matter = Matter that is not visible, but whose existence is inferred from its gravitational effects on visible matter, *radiation*, and the large-scale structure of the Universe.

Degree of freedom = Any of a number of parameters within a system that may vary independently, e.g., one of the vibrational modes of a molecular bond.

Dobson unit = A unit of measurement of the density of a trace gas (usually ozone) in a vertical column of Earth's atmosphere. One Dobson unit is equivalent to a layer of the pure gas 10 micrometers thick under standard *temperature* and pressure at Earth's surface.

Doppler effect = An increase or a decrease in the frequency of a sound wave from a source moving toward or away from the receiver, respectively. Also an increase or a decrease in the frequency of light (blueshift or redshift) as an object moves toward or away from the viewer, respectively.

Dry fog = A fog that occurs above the dew point due to the binding of water vapor to a hygroscopic (water-attracting) *aerosol*, such as *sulfuric acid*. Reported in Europe during the summer following the 1783 eruption of Laki *volcano* in Iceland.

Eccentricity = Deviation of a planetary orbit from circularity.

Eemian climatic optimum = The interglacial stage of the Pleistocene epoch preceding the most recent (Weichsel) glacial stage, from about 130,000 to about 115,000 years ago.

Effusive volcanism = A style of volcanic eruption, typical today in Hawaii and Iceland, in which highly fluid, high-temperature *magma* rises to Earth's surface and emerges as lava fountains feeding lava flows that move relatively quietly outward from the vent across the land surface. Effusive volcanic eruptions deplete *ozone*, causing warming, but rarely explode enough material into the lower *stratosphere* to form *aerosols* in sufficient quantities to reflect and scatter sunlight, thereby causing net cooling. Effusive volcanoes erupt over intervals from days to hundreds of thousands of years.

Electromagnetic radiation = Radiation emitted by any body of matter and thought of as an electric field, whose changes with time induce a magnetic field, whose changes in time induce an electric field, and so on, appearing to propagate at the speed of light. The spectrum of electromagnetic radiation extends from low frequency radio signals, to visible light, to ultra-high-frequency gamma rays (Figure 4.3 on page 48).

Electronic transition = The sudden motion of an electron from one energy level within a lower orbital of an atom to a higher energy level in a higher orbital of the atom upon absorption of electromagnetic energy whose frequency exceeds a certain threshold. Also, the sudden, spontaneous return of an electron from a higher energy level to a lower energy level with the release of a *quantum* of energy.

Energy = What makes things happen. The "go" of the Universe. A physical property of objects that can be transferred to other objects and converted into different forms, but cannot be created or destroyed.

Entropy = A measure of disorder in a thermodynamic system.

Exoelectrons = Weak electrons emitted during rock fracture, leading to the formation of *ozone*. A possible explanation for large observed emissions of ozone days to months before volcanic eruptions as rising *magma* fractures overlying bedrock.

Explosive volcanism = A style of volcanic eruption in which viscous, low-temperature *magma* is ejected violently from a shallow, subsurface chamber, producing an eruption column of gas and *ash* that often rises more than 19 miles (30 km)—well into the *stratosphere*. Explosive volcanoes typically eject megatons of water and *sulfur dioxide* into the lower stratosphere, where they form *sulfuric acid aerosols*, whose particle sizes grow large enough to reflect and scatter *ultraviolet radiation* from Sun, thereby cooling Earth. *Explosive* volcanoes erupt for periods of hours to days.

Extrusive volcanism = A volcano that extrudes lava flows out on the ground, usually without large explosions. Another term for *effusive volcanism*.

Fluorescence = The emission of monochromatic *electromagnetic radiation* due to excited electrons falling back to lower energy states in a substance that has been irradiated by radiation of higher frequency.

Foraminifera = A class of unicellular marine microorganisms, with or without calcium carbonate tests, or shells. Some species are planktonic (floating on the sea surface), whereas others are benthic (living on the sea floor).

Forbidden transition = The raising of an electron to a higher energy state, or its falling back to a lower energy state, in violation of *quantum mechanical* selection rules. An improbable, but not impossible, occurrence.

Frequency of oscillation = The number of times per second that an oscillator completes an oscillation.

Fumarole = A hole in the ground near volcanoes from which gases and steam arise.

GISP2 = The 3053-m-deep Greenland Ice Sheet Project Borehole Number 2, located beneath Summit Camp, Greenland.

Gravitational redshift = A slight reduction in the frequency of *electromagnetic radiation* leaving a strong gravitational field source.

Greenhouse gas = Gas molecules containing 3 or more atoms that absorb infrared energy and are thought by some to cause global warming. The primary greenhouse gases in Earth's atmosphere are water vapor, carbon dioxide, methane, nitrous oxide, and ozone.

Halogen = Any of the five monovalent, nonmetallic elements, fluorine, chlorine, bromine, iodine, and astatine, in group 17 of the periodic table of the elements. Chlorine and bromine are the primary cause of *ozone* depletion.

Heinrich event = Any of 10 events, observed in sea-floor sediments between 115,000 and 15,000 years ago, in which large armadas of icebergs broke off from glaciers and drifted across the North Atlantic Ocean, dropping entrained boulders to the sea floor as they melted.

Ignimbrite = A consolidated or heat-welded deposit of volcanic *ash*, glass shards, crystal fragments, and rock fragments laid down by a nuée ardente, or *pyroclastic* flow—a sizzling, incandescent avalanche of intermediate to *silicic magma*, ejected from a volcanic vent during an eruption. An ignimbrite filled the Valley of 10,000 Smokes during the eruption of Mount Katmai/Novarupta in 1912.

Infrared radiation = Electromagnetic *radiation* with *frequencies* below *visible* light in the range from 0.3 to 430 THz (*terahertz*) and *energies* less than 1.7 eV (electron volts).

Internal energy = The thermal energy stored in matter in the form of random thermal motions of molecules, bond vibrations and rotations, and electron excitations. Internal energy increases with increasing *temperature* of matter. The amount of energy that can be stored—the heat capacity—increases with the number of *degrees of freedom* of these thermal motions.

Intrusive = A coarse-grained, crystalline, igneous rock solidified from *magma* emplaced below Earth's surface.

Ionosphere = A region of Earth's upper atmosphere at altitudes from about 37 mi (60 km) up to 620 mi (1000 km), in which solar *radiation* ionizes most chemical species, including nitrogen and oxygen, which make up more than 99% of Earth's atmosphere.

Island arc = A typically arcuate chain of islands, often bearing *andesitic volcanoes* and located on the overriding *tectonic plate* above a *subduction zone*.

Kinetic energy = A form of energy attributed to a material body by virtue of its motion and equal to one-half its mass times the square of its velocity.

Last glacial maximum (LGM) = The time, 26,500 years ago, when the ice sheets of the most recent glacial stage reached their maximum extension.

Lithospheric plate = See *Tectonic plate*.

Luminiferous aether = The hypothetical material medium in which light was thought to propagate through space as a wave. Postulated by Fresnel in 1818 and proved not to exist by Michelson and Morley in 1887.

Magma = Molten and semi-molten subsurface rock containing solids, suspended crystals, dissolved gases, and, if vapor pressure exceeds confining pressure, gas bubbles.

Magma chamber = A large, subterranean cavity filled with *magma*.

Mantle = A thick shell of ultramafic rock (rock containing less than 46% silicon dioxide) in Earth's interior, extending down from the Mohorovičić discontinuity at the base of Earth's crust—located at a depth ranging from 3.1 mi (5 km) beneath oceans to 50 mi (80 km) beneath continents—to the outer boundary of Earth's core at a depth of 1800 mi (2885 km).

Mesopause = The coldest region of Earth's atmosphere at altitudes from 52 to 62 mi (85 to 100 km), where the rate of cooling with increasing height decreases to zero. Above the *mesopause* temperatures increase with height.

Mesosphere = A region of Earth's upper atmosphere lying above the *stratopause*, at about 30 mi (48 km) altitude, and below the *mesopause*, at about 50 mi (80 km) altitude, in which *temperature* decreases upward and turbulence occurs.

Milanković cycle = Any of the three cyclic variations in Earth's orbit around Sun, involving the obliquity of Earth's rotational axis, the *precession* of the equinoxes, and the *eccentricity* of Earth's orbit.

Montreal Protocol = The Montreal Protocol on Substances that Deplete the Ozone Layer is a protocol under the Vienna Convention for the Protection of the Ozone Layer. It is an international treaty aimed at protecting the *ozone layer* by phasing out the production of *chlorofluorocarbons* (*CFCs*) and other ozone depleting substances. It was first agreed to on in September 1987, entered into force in January 1989, and has been revised eight times.

Oceanic plateau = A large, relatively flat, *submarine* region that rises well above the level of the ambient seabed. Many are remnants of large igneous provinces formed by the marine equivalent of continental flood *basalts*.

Oceanic ridge = A continuous, 52,000 mile (84,000 km) long, *submarine* mountain chain traversing ocean floors, along the axis of which seafloor spreading occurs. The ridge typically rises 0.6 to 1.9 mi (1 to 3 km) above the ocean floor and is offset by hundreds of transverse fracture zones.

Oxygen fugacity = The *partial pressure* of oxygen gas in a high-pressure system, corrected for non-ideal gas behavior.

Oxygen isotope = Any of three naturally occurring versions of the oxygen atom, differing only in the number of neutrons in the nucleus. Most abundant in Earth's atmosphere is ^{16}O at 99.759%; next is ^{18}O at 0.204%; and least abundant is ^{17}O at 0.037%.

Ozone = A molecule made up of three atoms of oxygen (O_3).

Ozone layer = A region extending primarily from altitudes of 12 to 19 miles (20 to 30 km) in the lower stratosphere where ozone is continually formed and dissociated by high-energy ultraviolet solar radiation, heating the lower stratosphere. Without the ozone layer, Earth would not have a stratosphere.

Paleocene-Eocene Thermal Maximum (PETM) = A dramatic global warming event at about 55.8 Ma (million years ago) and lasting about 170,000 years, during which *temperature* rose about 5°C as the separation of Greenland from Europe generated the massive, *basaltic* eruptions of the North Atlantic Igneous Province.

Paleotemperatures = *Temperatures* in the geologic past, typically reconstructed with the aid of *proxy* records, such as $\delta^{18}O$, or tree-ring time series.

Paleotropical = Referring to the tropical portions of the world in the geologic past. (Also, in biogeography, refers to the Old World tropics.)

Partial pressure = The hypothetical pressure of a gas if that gas alone occupied the same volume as a mixture of gases, including the one in question. Partial pressure is a measure of the thermodynamic activity of the gas molecules.

Parts per billion (ppb) = The number of molecules of a specific chemical species in a mixed population of one billion molecules.

Parts per million (ppm) = The number of molecules of a specific chemical species in a mixed population of one million molecules.

***p*CO2** = The *partial pressure* of *carbon dioxide*.

Photochemistry = The branch of chemistry concerned with the chemical effects of electromagnetic *radiation* (light). Also, chemical reactions caused by absorbing electromagnetic radiation.

Photodissociation = The breaking of a molecular bond on absorption of electromagnetic *radiation* of sufficiently high frequency.

Photoionization = The removal of an electron from a molecule, atom, or negative ion by high-frequency ultraviolet *radiation* in a planetary *ionosphere*. No molecular bonds are broken in photoionization. See people.virginia.edu/~rej/MAE494/Part-4-07.pdf.

Photon = In modern physics, the photon is considered to be an elementary particle that comprises a *quantum* of *electromagnetic radiation* (light), whose energy content varies with its frequency. In this book, I suggest that the photon is a very useful mathematical variable, but is not a physical thing.

Plate = See *Tectonic plate*.

Planck's law = An empirical formula that describes the *electromagnetic radiation* emitted by a black body (a perfect absorber and emitter of radiation) in *thermal equilibrium*.

Planck constant ($h=E/v$) = The amount of energy produced in one second by *electromagnetic radiation* of a given frequency.

Planck-Einstein relation ($E=hv$) = The statement that the energy of an atomic oscillator is equal to the Planck constant times the *frequency of oscillation*.

Polar jet stream = A fast-flowing, narrow air current near the altitude of the *tropopause* flowing from west to east, typically just outside of the *polar vortex*, and with a meandering path consisting of ridges and troughs (Rossby waves), which directly affect weather.

Polar night jet = A jet stream that forms only during winter months at latitudes around 60° and at altitudes around 30 mi (48 km).

Polar stratospheric cloud (PSC) = A thin cloud type formed in polar regions at altitudes between 9 and 15.5 mi (15to 25 km) and at *temperatures* below -108°F (-78°C).

PSCs support chemical reactions that produce active chlorine, which catalyzes ozone depletion.

Polar vortex = A persistent, large-scale area of closed, circular air motion circling a polar region and extending up from the middle to upper *troposphere* into the *stratosphere*. This cold-core, low-pressure vortex is stronger in the winter and is enhanced by *ozone* depletion.

Pollution = Human generated harmful or toxic waste in quantities greater than what can be degraded locally by Earth's natural recycling processes.

Potential energy = The energy that an object has due to its position in a gravitational or electromagnetic force field.

Pre-boreal warming = A rapid global warming event 10,300 years ago, continuing more gradually to 9500 years ago, and ending the last ice age.

Precession (Axial) = A circular path, described by the north pole as Earth's rotational axis shifts westward through 1° every 72 years due to gravitational forces from Moon and Sun exerted on Earth's equatorial bulge, which maintain the axial inclination at 23.4 ±1.25° from a perpendicular to the orbital plane.

Proxy = A variable that can be measured to infer the value of another variable that cannot be measured directly. The ratio of *oxygen isotopes* ($\delta^{18}O$) or the thicknesses of tree rings are often used as proxies for *temperature*. They can be used to infer the trends in temperatures over long periods of time.

Pumice = Glassy volcanic rock consisting mainly of vesicles, or frozen bubbles, formed by escaping gas. Pumice typically floats on water.

Pyroclastic rock = Rock composed only or primarily of consolidated fragments of *tephra*.

Quantum (plural quanta) = The minimum amount of any physical entity involved in an interaction in which the entity is assumed to have values that are only of certain discrete magnitudes.

Quantum mechanics = The study of interactions between matter and energy at scales typical of atoms and subatomic particles, at which scales the microstructures of matter constrain energy transfers to discrete steps, or *quanta*.

Radiation = The emission or transmission of thermal energy through air or space in the form of frequency (color) and *amplitude* (intensity).

Radiative forcing = The difference between *radiation* from Sun absorbed by Earth and thermal energy radiated by Earth back into space, typically quantified at the *tropopause* in watts per square meter.

Resonance = A phenomenon that occurs when a given oscillating system, such as a molecular bond, is driven by another oscillating system at the same frequency, causing some of the *amplitude* of oscillation from the higher amplitude system to be shared, increasing the amplitude of oscillation of the lower amplitude system.

Rhyolite = A highly evolved (has assimilated much crustal material), fine-grained, usually light-colored *extrusive* volcanic rock consisting mostly of potassium- and sodium-rich plagioclase feldspars (potassium and sodium aluminum silicates) and quartz with minor muscovite and biotite mica and/or amphibole. The fine-grained equivalent of granite.

Seamount = A *submarine volcano* that rises at least 1000 m above the sea floor.

Silicic magma = *Magma* that contains over 63% silicon dioxide.

Sill = A thin, flat, *intrusive* body of igneous rock, the borders of which are predominantly parallel with the stratification or foliation in the surrounding bedrock.

Spreading rift (a.k.a. Spreading center) = A boundary between two adjacent, diverging *tectonic plates*, along which new material is added to their trailing margins by the upwelling of *magma* within the median rift.

Stratopause = The upper limit of the *stratosphere*, at an altitude of about 31 to 34 mi (50 to 55 km), above which *temperature* decreases with altitude.

Stratosphere = A stable, stratified, region of Earth's atmosphere, in which *temperature* increases upward due to heat produced by the *photodissociation* of oxygen, *ozone*, and other chemical species in the Chapman cycle. The base of the stratosphere is at the *tropopause,* which varies from 5.6 mi (9 km) altitude at the poles to 11 mi (17 km) at the equator. Its top is at the *stratopause* at an altitude of 31 to 34 mi (50 to 55 km). The stratosphere is a product of Earth's thermally active *ozone layer*, without which it could not exist.

Subaerial volcanic eruptions = Eruptions emitting volcanic gases and *tephra* directly into the atmosphere, contrasted with *submarine volcanic* eruptions, from which most gases do not reach the atmosphere but do change the chemistry of sea water.

Subduction zone = A region where one *lithospheric plate* dips gently to steeply beneath the edge of another, converging *lithospheric plate* and descends into Earth's *mantle*.

Submarine volcanic eruptions = Eruptions on the sea floor from which *tephra* and most volcanic gases are not emitted directly into the atmosphere but can change the chemistry of sea water. Contrasted with *subaerial* volcanic eruptions, in which *tephra* and gases do enter the atmosphere.

Sulfate = A doubly negatively charged anion (SO_4^{2-}) typically formed by the oxidation and hydration of *sulfur dioxide* (SO_2), which is erupted by *volcanoes*—its major source. Sulfate concentrations measured in ice cores provide a *proxy* for volcanic activity—a measure of the amount of sulfur dioxide emitted by volcanoes during the time when a particular ice layer sat on the surface as snow. Sulfate is the second most common anion in the ocean after chloride.

Sulfur dioxide = A pungent, toxic, colorless gas consisting of molecules of a compound composed of one atom of sulfur bound asymmetrically to two atoms of oxygen. Sulfur dioxide is the third most voluminous gas emitted from *volcanoes* after water and *carbon dioxide*.

Sulfuric acid = A powerful acid formed from *sulfur dioxide* and water in the lower *stratosphere* after major *explosive* volcanic eruptions. It has low vapor pressure, which allows it to persist, clinging to particles in the air. In the stable, stratified environment of the lower *stratosphere*, sulfuric acid particles increase in size until they become large enough to reflect and scatter *ultraviolet* and *visible* light, thereby reducing the amount of solar energy that reaches Earth's surface, thus cooling Earth.

Table mountain (tuya) = A flat-topped mountain formed by the *extrusion* of *basaltic* lava beneath a thick cover of glacier ice, which cools the lava rapidly, retarding its lateral spread. The tuya is revealed upon melting of the superjacent ice.

Tectonic plate (lithospheric plate) = A rigid slab of oceanic crust and underlying upper *mantle* with or without a thick, uppermost layer of continental crust. Oceanic lithospheric plates are about 62 mi (100 km) thick, while continental plates are closer to 120 mi (200 km) thick. Each moves as an independent unit on the underlying, partially molten asthenosphere, or low velocity zone.

Temperature = A measure of the average *kinetic* energy in the microscopic motions of the constituent particles of matter, such as electrons, atoms, and molecules. Temperature of an ideal gas is proportional to the average kinetic energy of all the molecules making up the gas, which is proportional to the average velocity of the molecules squared. In matter, temperature is the result of oscillations of the chemical bonds that hold matter together. The higher the *frequencies* and *amplitudes* of oscillation, the higher the temperature.

Tephra = Fragmental volcanic material produced during a volcanic eruption.

Terahertz (THz) = An oscillation frequency of one trillion (10^{12}) cycles per second.

Thermal equilibrium = A condition in which the distribution of heat in a body of matter is homogeneous, and no transfer of heat is taking place within that body.

Thermodynamics = The study of heat in motion.

Thermosphere = The layer of Earth's atmosphere above the *mesopause*, at about 53 mi (85 km) altitude, within which *ultraviolet radiation photoionizes* and *photodissociates* molecules of nitrogen and oxygen. The most energetic solar *radiation* is absorbed near the top of the thermosphere, causing the highest atmospheric temperatures, although the air is so thin at those altitudes that measured temperatures are very cold.

Toba = One of the largest volcanoes in the world in northern Sumatra, Indonesia. Toba exploded in a colossal eruption (*VEI* 8) about 73,000 years ago, causing a volcanic winter with a worldwide decrease in *temperature* of between 5.4 and 9.0°F (3.0 and 5.0°C), and of up to 27.0°F (15.0°C) in higher latitudes. Today, Toba's *caldera* contains a large lake.

Trachyandesite = An evolved (has assimilated much crustal material), *extrusive* igneous rock with little or no free quartz, dominated by alkali feldspar and sodic plagioclase (potassium and sodium aluminum silicates) along with minor amounts of amphibole, biotite, or pyroxene.

Transform fault = A vertical zone of relative motion between the margins of juxtaposed sections of two *lithospheric plates* spreading in opposite directions from offset segments of a median rift, which terminate both ends of the fault zone.

Tropopause = The upper surface of the *troposphere*, located at an altitude of about 5.6 mi (9 km) in polar regions and at about 11 mi (18 km) over the equator. Local changes in total column *ozone* cause changes in the height of the tropopause.

Troposphere = The shell of the atmosphere adjacent to Earth's surface, characterized by thermal instability due to a decrease in temperature with altitude. The troposphere extends upward from Earth's surface to 5.6 mi (9 km) in polar regions and at about 11 mi (18 km) over the equator.

Turbopause = The altitude around 62 mi (100 km) that is the boundary between well-mixed chemical gas species below that display identical height distributions and the region above where molecular diffusion dominates causing the chemical compositions to vary by chemical species.

Tuya = See *Table mountain*

Ultraviolet radiation = *Electromagnetic radiation* with *frequencies* in the range from 790 to 30,000 THz and *energies* from 3.3 to 124 eV (electron volts).

Visible radiation = *Electromagnetic radiation* with *frequencies* in the range from 430 to 790 THz and *energies* from 1.7 to 3.3 eV (electron volts).

Volcanic Explosivity Index (VEI) = An arbitrary, logarithmic scale, ranging from 0 to 8, of the intensity of volcanic eruptions, based on volume of ejecta, eruption cloud height, and subjective factors. VEI 6 represents a large eruption such as Mt. Pinatubo in 1991. VEI 8 represents a gigantic eruption such as *Toba*, 73,000 years ago.

Volcano = A constructional landform consisting of solidified and/or pulverized *magmatic* materials (lava and *tephra*) ejected from a central vent onto Earth's surface.

Wave-particle duality = The consideration in *quantum mechanics* that light is best described sometimes as waves and at other times as particles.

Welded tuff = See *Ignimbrite*.

ABOUT THE AUTHOR

Peter Langdon Ward received a BA from Dartmouth College in Geology in 1965 and a PhD from Columbia University in Geophysics in 1970. His thesis focused on earthquakes, volcanoes, and tectonics in Iceland. He worked for 27 years with the United States Geological Survey in Menlo Park, California, as a geophysicist, as a Program Manager, and as Chief of the Branch of Seismology, which, under his leadership, became the Branch of Earthquake Mechanics and Prediction. He played a major role in developing and leading the National Earthquake Hazard Reduction Program.

He spent 8 summers studying earthquakes and volcanoes in Katmai National Park, Alaska, and has installed instruments on volcanoes in Iceland, Alaska, Hawaii, Washington State, California, Guatemala, El Salvador, and Nicaragua.

At the age of 31, he was appointed to lead a group of 40 PhD scientists and 100 other workers in a major effort to monitor earthquakes, with the goal of predicting their time of occurrence. This led to life-long interests in how to manage creativity—an obvious oxymoron—and in the public policy aspects of good science.

He specialized for many years in helping the general public understand the risks they face from natural hazards so that families could take reasonable actions to live more safely. He set a new standard, in 1990, by conceiving, creating, finding funding for, producing, and distributing a 24-page magazine, with versions in English, Chinese, Spanish, and Braille, to 3.3 million families in Northern California explaining future earthquake risk and simple actions that everyone can take to be prepared. His effort was featured on Good Morning America, and he received the Public Affairs Award from the Secretary of the Interior, the highest award of the Association of Government Communicators, and was a Finalist for Federal Employee of the Year Award in 1991.

Peter feels that his greatest contribution in life was raising four children. His family was featured in the New York Times Sunday Magazine and on the Phil Donahue Show as being a very successfully blended marriage involving his two children from a previous marriage and his wife's two children, also from a previous marriage. He has six grandchildren.

Peter has played classical and folk piano all his life. He also played accordion, leading a Balkan-Bulgarian folk dance band for several years, and he especially loves leading group singing of folk songs with his 12-string guitar. From an early age, he enjoyed camping, canoeing, hiking, skiing, and especially mountain climbing in both summer and winter. He has rowed his own raft through the Grand Canyon twice.

Peter retired to Jackson Hole, Wyoming, with his wife, to enjoy Nature, but he has been obsessed since 2006 with trying to understand what really causes global warming.

END NOTES

Preface

1   en.wikipedia.org/wiki/Open-source_software

2   nature.com/nature/journal/v438/n7070/full/438900a.html

Overview

3   nobelprize.org/nobel_prizes/chemistry/laureates/1995/

4   nobelprize.org/nobel_prizes/chemistry/laureates/1995/molina-lecture.html, page 1

5   en.wikipedia.org/wiki/John_Tyndall

6   Photo © Arctic-Images/Corbis

7   en.wikipedia.org/wiki/Holuhraun

8   en.wikipedia.org/wiki/Laki and en.wikipedia.org/wiki/Causes_of_the_French_Revolution

9   scienceandpublicpolicy.org/commentaries_essays/crichton_three_speeches.html, page 6

Chapter 1

10   Griggs, R. F., 1922, The Valley of Ten Thousand Smokes, National Geographic Society, 341 p., page 191

11   Ward, P., and Matumoto, T., 1967, A summary of volcanic and seismic activity in Katmai National Monument, Alaska: Bulletin Volcanologique, v. 31, no. 1, p. 107-129, doi:10.1007/BF025970090.1007/BF02597009

12   Castle, R. O., and Gilmore, T. D., 1992, A revised configuration of the southern California uplift: Geological Society of America Bulletin, v. 104, no. 12, p. 1577-1591, doi:10.1130/0016-7606(1992)104<1577:ARCOTS>2.3.CO;2

13   livingmoresafely.com/earthquakes.html

14   hvo.wr.usgs.gov/

15   youtube.com/watch?v=KNpdwd53qgM

16   Griggs, R. F., 1922, The Valley of Ten Thousand Smokes, National Geographic Society, 341 p., page 33

17   Ward, P. L., 2009, Sulfur dioxide initiates global climate change in four ways: Thin Solid Films, v. 517, no. 11, p. 3188-3203, doi:10.1016/j.tsf.2009.01.005

18   Staehelin, J., et al., 1998, Trend analysis of the homogenized total ozone series of Arosa (Switzerland), 1926–1996: Journal of Geophysical Research, v. 103, no. D7, p. 8389-8399, doi:10.1029/97JD03650

Chapter 2

19   rci.rutgers.edu/~schlisch/103web/Pangeabreakup/extinctions.html

20   en.wikipedia.org/wiki/Permian-Triassic_extinction_event

21   Reichow, M. K., et al., 2009, The timing and extent of the eruption of the Siberian Traps large igneous province: Implications for the end-Permian environmental crisis: Earth and Planetary Science Letters, v. 277, no. 1, p. 9-20, doi:10.1016/j.epsl.2008.09.030

22   en.wikipedia.org/wiki/Large_igneous_province

23   news.brown.edu/articles/2014/09/extinctions

24   NOAA, 2015, ncdc.noaa.gov/cag/time-series/global

25   Hughes, G. L., et al., 2007, Statistical analysis and time-series models for minimum/maximum temperatures in the Antarctic Peninsula: Proceedings of the Royal Society A: Mathematical, Physical and Engineering Science, v. 463, no. 2077, p. 241-259

26   en.wikipedia.org/wiki/Current_sea_level_rise

27   nbcnews.com/science/environment/californias-drought-worst-1-200-years-researchers-say-n262621. Griffin, D., and Anchukaitis, K. J., 2014, How unusual is the 2012–2014 California drought?: Geophysical Research Letters, v. 41, no. 24, p. 9017-9023, doi:10.1002/2014GL062433

28   keepcaliforniafarming.org/california-water-crisis/farming-california-news/if-we-didnt-have-californias-food-what-would-we-eat/

29   pewinternet.org/files/2015/01/PI_ScienceandSociety_Report_012915.pdf

30   Mermin, N. D., 2004, Could Feynman have said this?: Physics Today, v. 57, no. 5, p. 10, doi:10.1063/1.1768652

31   marxists.org/reference/subject/philosophy/works/dk/bohr.htm

32   Molina, M. J., and Rowland, F. S., 1974, Stratospheric sink for chlorofluoromethanes: Chlorine catalysed destruction of ozone: Nature, v. 249, p. 810-814

33   Farman, J. C., et al.,1985, Large losses of total O_3 in atmosphere reveal seasonal ClO_x/NO_x interaction: Nature, v. 315, p. 207-210

34   gisp2.sr.unh.edu

Chapter 3

35   en.wikipedia.org/wiki/Temperature_measurement

36   en.wikipedia.org/wiki/Central_England_temperature

37   cru.uea.ac.uk/cru/data/temperature/crutem4/station-data.htm

38   Jones, P. D., et al., 1999, Surface air temperature and its changes over the past 150 years: Reviews of Geophysics, v. 37, no. 2, p. 173–199, doi:10.1029/1999RG900002

39   cru.uea.ac.uk/cru/data/temperature/

40   en.wikipedia.org/wiki/HadCRUT

41   cru.uea.ac.uk/cru/data/temperature/#faq5

42   cru.uea.ac.uk/cru/data/temperature/HadCRUT4.png

43   en.wikipedia.org/wiki/Medieval_Warm_Period

44   en.wikipedia.org/wiki/Little_Ice_Age

45   Hansen, J., and Lebedeff, S., 1987, Global trends of measured surface air temperature: Journal of Geophysical Research, v. 92, no. 13, p. 13345-13372, doi:10.1029/JD092iD11p13345

46   data.giss.nasa.gov/gistemp/

47   ncdc.noaa.gov/data-access/land-based-station-data/land-based-datasets/global-historical-climatology-network-ghcn

48   coads.noaa.gov/

49   en.wikipedia.org/wiki/Berkeley_Earth

50 Rohde, R., et al., 2013, A new estimate of the average earth surface land temperature spanning 1753 to 2011: Geoinformatics & Geostatistics: An Overview, v. 1, no. 1, p. 7, doi:10.4172/2327-4581.1000101

51 Karl, T. R., et al., 2015, Possible artifacts of data biases in the recent global surface warming hiatus: Science, v. 348, no. 6242, p. 1469-1472 doi:10.1126/science.aaa5632

52 Arrhenius, S., 1896, On the influence of carbonic acid in the air upon the temperature of the ground: Philosophical Magazine and Journal of Science Series 5, v. 41, no. 251, p. 237-276, doi:10.1080/14786449608620846

53 PALAEOSENS Project Members, 2012, Making sense of palaeoclimate sensitivity: Nature, v. 491, no. 7426, p. 683-691, doi:10.1038/nature11574

54 en.wikipedia.org/wiki/IPCC_Fifth_Assessment_Report

55 esrl.noaa.gov/gmd/ccgg/trends/#mlo_full

56 cdiac.ornl.gov/ftp/trends/co2/lawdome.combined.dat

57 ozonedepletiontheory.info/gg-warming-hiatus.html

58 Kerr, R. A., 2009, What happened to global warming? Scientists say just wait a bit: Science, v. 326, no. 2 October 2009, p. 28-29, doi:10.1126/science.326_28a

59 Solomon, S., 1999, Stratospheric ozone depletion: A review of concepts and history: Reviews of Geophysics, v. 37, no. 3, p. 275-316, doi:10.1029/1999RG900008

60 earthobservatory.nasa.gov/Features/WorldOfChange/ozone.php

61 Molina, M. J., and Rowland, F. S., 1974, Stratospheric sink for chlorofluoromethanes: Chlorine catalysed destruction of ozone: Nature, v. 249, p. 810-814, doi:10.1038/249810a0

62 Farman, J. C., et al., 1985, Large losses of total O_3 in atmosphere reveal seasonal ClO_x/NO_x interaction: Nature, v. 315, p. 207-210, doi:10.1038/315207a0

63 Staehelin, J., et al., 1998, Total ozone series at Arosa (Switzerland): Homogenization and data comparison: Journal of Geophysical Research, v. 103, no. D5, p. 5827-5841, doi:10.1029/97JD02402

64 Levitus, S., et al., 2012, World ocean heat content and thermosteric sea level change (0-2000 m), 1955-2010: Geophysical Research Letters, v. 39, no. 10, p. L10603, doi:10.1029/2012GL051106

65 Tedetti, M., and Sempéré, R., 2006, Penetration of ultraviolet radiation in the marine environment. A review: Photochemistry and Photobiology, v. 82, no. 2, p. 389-397, doi:10.1562/2005-11-09-IR-733

66   Hughes, G. L., et al., 2007, Statistical analysis and time-series models for minimum/maximum temperatures in the Antarctic Peninsula: Proceedings of the Royal Society A: Mathematical, Physical and Engineering Science, v. 463, no. 2077, p. 241-259, doi:10.1098/rspa.2006.1766

67   Mulvaney, R., et al., 2012, Recent Antarctic Peninsula warming relative to Holocene climate and ice-shelf history: Nature, v. 489, no. 7414, p. 141-144, doi:10.1038/nature11391

68   Bromwich, D. H., et al., 2013, Central West Antarctica among the most rapidly warming regions on Earth: Nature Geoscience, v. 6, p. 139-145, doi:10.1038/ngeo1671

69   Trenberth, K. E., et al., 2007, Observations: Surface and Atmospheric Climate Change, in Solomon, S., et al., eds., Climate Change 2007: The Physical Science Basis. Contribution of Working Group I to the Fourth Assessment Report of the Intergovernmental Panel on Climate Change, Cambridge University Press, pages 235-336

70   en.wikipedia.org/wiki/Polar_amplification

71   Serreze, M. C., and Barry, R. G., 2011, Processes and impacts of Arctic amplification: A research synthesis: Global and Planetary Change, v. 77, no. 1, p. 85-96, doi:10.1016/j.gloplacha.2011.03.004

Chapter 4

72   Fresnel, A., Lettre d'Augustin Fresnel à François Arago sur l'influence du mouvement terrestre dans quelques phénomènes d'optique, in Proceedings Annales de chimie et de physique1818, Volume 9, p. 57-66

73   Faraday, M., 1849, The bakerian lecture: Experimental researches in electricity. twenty-second series: Philosophical Transactions of the Royal Society of London, v. 139, p. 1-18, doi:10.1098/rstl.1849.0001

74   ffden-2.phys.uaf.edu/webproj/212_spring_2014/Amanda_Mcpherson/Amanda_McPherson/em_electric_magnetic_propagating_waves.jpg

75   Maxwell, J. C., 1865, A dynamical theory of the electromagnetic field: Philosophical Transactions of the Royal Society of London, v. 155, p. 459-512, doi:10.1098/rstl.1865.0008

76   en.wikipedia.org/wiki/Timeline_of_luminiferous_aether

77   Michelson, A. A., and Morley, E. W., 1887, On the relative motion of the earth and the luminiferous ether: American journal of science, v. 34, no. 203, p. 333-345, doi:10.2475/ajs.s3-34.203.333

78   en.wikisource.org/wiki/A_Heuristic_Model_of_the_Creation_and_
      Transformation_of_Light

79   en.wikipedia.org/wiki/Photon

80   en.wikipedia.org/wiki/Wave-particle_duality

81   books.google.com/
      books?id=BT8AAAAAYAAJ&pg=PA242&hl=en#v=onepage&q&f=false

82   Maxwell, J. C., 1873, A treatise on electricity and magnetism, Oxford, Clarendon
      Press, 560 p.

83   Feynman, R. P., et al., 1963, Feynman Lectures on Physics. vol. 1: Mainly
      mechanics, radiation and heat, Addison-Wesley, page 4-2

84   Coopersmith, J., 2010, Energy, the Subtle Concept: The discovery of Feynman's
      blocks from Leibniz to Einstein, Oxford University Press 400 p., page 350

85   en.wikipedia.org/wiki/Conservation_of_energy

86   youtube.com/watch?v=BE827gwnnk4

87   youtube.com/watch?v=aCocQa2Bcuc

88   youtube.com/watch?v=nFzu6CNtqec

89   youtube.com/watch?v=VBssGPfYBr4

90   youtube.com/watch?v=mcjTUpEnMnw

91   youtube.com/watch?v=tI6S5CS-6JI

92   Grossman, J. C., 2014, Thermodynamics: Four laws that move the Universe, The
      Great Courses, Course 1291, transcript page 74

93   en.wikipedia.org/wiki/Molecular_vibration

94   en.wikipedia.org/wiki/Wien_approximation

95   Grossman, J. C., 2014, Thermodynamics: Four laws that move the Universe, The
      Great Courses, Course 1291, transcript page 49.

96   en.wikipedia.org/wiki/Planck%E2%80%93Einstein_relation

97   en.wikipedia.org/wiki/Wien_approximation

98   en.wikipedia.org/wiki/Solid_angle

99   Trenberth, K. E., and Fasullo, J. T., 2012, Tracking Earth's energy: From El
      Niño to global warming: Surveys in Geophysics, v. 33, no. 3-4, p. 413-426,
      doi:10.1007/s10712-011-9150-2.

100   en.wikipedia.org/wiki/Photoelectric_effect#20th_century

101   en.wikipedia.org/wiki/Photoelectric_effect

102   Rothman, L. S., et al., 2013, The HITRAN2012 molecular spectroscopic database:
      Journal of Quantitative Spectroscopy and Radiative Transfer, v. 130, p. 4-50,
      doi:10.1016/j.jqsrt.2013.07.002

103   en.wikipedia.org/wiki/Stratopause

104   en.wikipedia.org/wiki/Photoreceptor_cell#Function

Chapter 5

105   Committee on Extension to the Standard Atmosphere, 1976, U.S. Standard
      Atmosphere, 1976, Washington, D.C., U.S. Government Printing Office, 241
      p. everyspec.com/NASA/NASA-General/NASA_TM-X-74335_37294/. Also
      Krueger, A. J., and Minzner, R. A., 1976, A mid-latitude ozone model for the
      1976 US standard atmosphere: Journal of Geophysical Research, v. 81, no. 24, p.
      4477-4481, doi:10.1029/JC081i024p04477

106   DeMore, W., et al., 1997, Chemical kinetics and photochemical data for use in
      stratospheric modeling, Evaluation number 12: Pasadena, Jet Propulsion Lab.,
      California Inst. of Tech., jpldataeval.jpl.nasa.gov/pdf/Atmos97_Anotated.pdf. Page
      260, Figure 7

107   earth.huji.ac.il/data/pics/lecture 3 O2 absorption spectrum.pdf

108   Based on Figure 3.5 on page 74 of Liou, K. N., 2002, An Introduction to
      Atmospheric Radiation, Academic Press, 583p. The complex details are
      summarized at earth.huji.ac.il/data/pics/lecture 3 O2 absorption spectrum.pdf

109   Finlayson-Pitts, B. J., and Pitts, J. N., 1999, Chemistry of the Upper and Lower
      Atmosphere: Theory, Experiments, and Applications, San Diego, Academic Press,
      969 p. Page 91

110   Madronich, S., 1993, Trends and predictions in global UV, in Chanin, M. L., ed.,
      The Role of the Stratosphere in Global Change, Volume 8, NATO ASI Series I:
      Global Environmental Change: Berlin, Springer-Verlag, p. 463-471

111   Finlayson-Pitts, B. J., and Pitts, J. N., 1999, Chemistry of the Upper and Lower
      Atmosphere: Theory, Experiments, and Applications, San Diego, Academic Press,
      969 p., Tables 3-7, 3-15, 3-16, 3-17

112   France, J., et al., 2012, A climatology of stratopause temperature and height in the
      polar vortex and anticyclones: Journal of Geophysical Research: Atmospheres, v.
      117, no. D6, doi:10.1029/2011JD016893

113   en.wikipedia.org/wiki/Stratopause

Chapter 6

114   From Figure 1 in Fioletov, V., 2008, Ozone climatology, trends, and substances
      that control ozone: Atmosphere-Ocean, v. 46, no. 1, p. 39-67, doi:10.3137/
      ao.460103

115   en.wikipedia.org/wiki/Ozone-oxygen_cycle

116   en.wikipedia.org/wiki/Ozone_layer

117   Chapman, S., 1930, A theory of upper-atmospheric ozone: Memoirs of the Royal Meteorological Society, v. 3, no. 26, p. 103-125, rmets.org/sites/default/files/chapman-memoirs.pdf

118   en.wikipedia.org/wiki/Ozone_layer#/media/File:Ozone_cycle.svg

119   en.wikipedia.org/wiki/Dobson_unit

120   woudc.org

121   From Figure 2 in Fioletov, V., 2008, Ozone climatology, trends, and substances that control ozone: Atmosphere-Ocean, v. 46, no. 1, p. 39-67, doi:10.3137/ao.460103

122   Molina, M. J., and Rowland, F. S., 1974, Stratospheric sink for chlorofluoromethanes: Chlorine catalysed destruction of ozone: Nature, v. 249, p. 810-814, doi:10.1038/249810a0

123   Lu, Q.-B., 2013, Cosmic-ray-driven reaction and greenhouse effect of halogenated molecules: Culprits for atmospheric ozone depletion and global climate change: International Journal of Modern Physics B, v. 0, no. 0, p. 1350073, doi:10.1142/S0217979213500732

124   en.wikipedia.org/wiki/Ozone_depletion

125   en.wikipedia.org/wiki/File:Jetcrosssection.jpg

126   Waugh, D. W., and Polvani, L. M., 2010, Stratospheric polar vortices: The Stratosphere: Dynamics, Transport, and Chemistry, Geophysical Monograph Series 190, p. 43-57, doi:10.1029/2009GM000887

127   Robock, A., 2000, Volcanic eruptions and climate: Reviews of Geophysics, v. 38, no. 2, p. 191-219, doi:10.1029/1998RG000054.

128   exp-studies.tor.ec.gc.ca/cgi-bin/selectMap

129   ozonedepletiontheory.info/ImagePages/eyjafjallajokull-ozone-animation.html

130   Figure 6 in Fioletov, V., 2008, Ozone climatology, trends, and substances that control ozone: Atmosphere-Ocean, v. 46, no. 1, p. 39-67, doi:10.3137/ao.460103

131   Figure 1, page 1 in Fowler, D., et al., 2008, Ground-level ozone in the 21st century: future trends, impacts and policy implications: Royal Society Science Policy Report, V. 15 no. 08, 148 p. Also uneplive.org/media/imgs/maps/air%20quality/Present-day%20surface%20ozone%20%28ppbv%29.png

132   epa.gov/airtrends/ozone.html

133   www-atm.physics.ox.ac.uk/user/barnett/ozoneconference/dobson.htm

134   Reed, R. J., 1950, The role of vertical motion in ozone-weather relationships: Journal of Meteorology, v. 7, p. 263-267, doi:10.1175/1520-0469(1950)007<0263:TROVMI>2.0.CO;2

Chapter 7

135   ozonewatch.gsfc.nasa.gov/monthly/SH.html

136   Farman, J. C., et al., 1985, Large losses of total O_3 in atmosphere reveal seasonal ClO_x/NO_x interaction: Nature, v. 315, p. 207-210, doi:10.1038/315207a0

137   ozonewatch.gsfc.nasa.gov/statistics/annual_data.html

138   ozonewatch.gsfc.nasa.gov/monthly/monthly_2006-09_SH.html

139   cru.uea.ac.uk/cru/data/temperature/HadCRUT4-sh.dat

140   Figure 1-27 in Harris, N. R. P., et al., 1995, 1. Ozone measurements, Scientific Assessment of Ozone Depletion: 1994, WMO/UNEP, esrl.noaa.gov/csd/assessments/ozone/1994/

141   Hassler, B., et al., 2011, Changes in the polar vortex: Effects on Antarctic total ozone observations at various stations: Geophysical Research Letters, v. 38, no. 1, p. L01805, doi:10.1029/2010GL045542

142   Waugh, D. W., and Polvani, L. M., 2010, Stratospheric polar vortices, in The Stratosphere: Dynamics, Transport, and Chemistry: Geophysical Monograph, v. 190, p. 43-57, doi:10.1029/2009GM000887

143   Hughes, G. L., et al., 2007, Statistical analysis and time-series models for minimum/maximum temperatures in the Antarctic Peninsula: Proceedings of the Royal Society A: Mathematical, Physical and Engineering Science, v. 463, no. 2077, p. 241-259, doi:10.1098/rspa.2006.1766

144   Mulvaney, R., et al., 2012, Recent Antarctic Peninsula warming relative to Holocene climate and ice-shelf history: Nature, v. 489, no. 7414, p. 141-144, doi:10.1038/nature11391

145   Bromwich, D. H., et al., 2013, Central West Antarctica among the most rapidly warming regions on Earth: Nature Geoscience, v. 6, p. 139-145, doi:10.1038/ngeo1671

146   Parkinson, C. L., and Cavalieri, D. J., 2012, Antarctic sea ice variability and trends, 1979-2010: The Cryosphere, v. 6, p. 871-880, doi:10.5194/tc-6-871-2012

147   Meredith, M. P., and King, J. C., 2005, Rapid climate change in the ocean west of the Antarctic Peninsula during the second half of the 20th century: Geophys. Res. Lett, v. 32, no. 19, p. L19604, doi:10.1029/2005GL024042

148 Clarke, A., et al., 2007, Climate change and the marine ecosystem of the western Antarctic Peninsula: Philosophical Transactions of the Royal Society B: Biological Sciences, v. 362, no. 1477, p. 149-166, doi:10.1098/rstb.2006.1958

149 Purkey, S. G., and Johnson, G. C., 2012, Global contraction of Antarctic bottom water between the 1980s and 2000s: Journal of Climate, v. 25, no. 17, p. 5830-5844, doi:10.1175/JCLI-D-11-00612.1

150 Waugh, D. W., et al., 2013, Recent changes in the ventilation of the southern oceans: science, v. 339, no. 6119, p. 568-570, doi:10.1126/science.1225411

151 Clarke, A., et al., 2007, Climate change and the marine ecosystem of the western Antarctic Peninsula: Philosophical Transactions of the Royal Society B: Biological Sciences, v. 362, no. 1477, p. 149-166, doi:10.1098/rstb.2006.1958

152 Stammerjohn, S. E., et al., 2008, Sea ice in the western Antarctic Peninsula region: Spatio-temporal variability from ecological and climate change perspectives: Deep Sea Research Part II: Topical Studies in Oceanography, v. 55, no. 18, p. 2041-2058, doi:10.1016/j.dsr2.2008.04.026

153 Trenberth, K. E., et al., 2007, Observations: Surface and Atmospheric Climate Change, in Solomon, S., et al., eds., Climate Change 2007: The Physical Science Basis. Contribution of Working Group I to the Fourth Assessment Report of the Intergovernmental Panel on Climate Change, Cambridge University Press, p. 235-336

154 Andersen, S., and Knudsen, B., 2006, The influence of polar vortex ozone depletion on NH mid-latitude ozone trends in spring: Atmospheric Chemistry and Physics, v. 6, no. 10, p. 2837-2845, doi:10.5194/acp-6-2837-2006

155 es-ee.tor.ec.gc.ca/e/ozone/normalozone.htm#nh

156 Manney, G. L., et al., 2011, Unprecedented Arctic ozone loss in 2011: Nature, v. 478, p. 469-475, doi:10.1038/nature10556

157 Lemke, P., et al., 2007, Observations: Changes in snow, ice, and frozen ground, in Solomon, S., et al., eds., Climate Change 2007: The physical science basis, Cambridge University Press, p. 337-383

158 Jeffries, M. O., and Richter-Menge, J. A., 2012, State of Climate in 2011, Chapter 5: The Arctic: Bulletin of the American Meteorological Society, v. 93, no. 7, p. S127-S147, ncdc.noaa.gov/bams-state-of-the-climate/2011.php

159 Tingley, M. P., and Huybers, P., 2013, Recent temperature extremes at high northern latitudes unprecedented in the past 600 years: Nature, v. 496, no. 7444, p. 201-205, doi:10.1038/nature11969

160   Comiso, J. C., 2006, Arctic warming signals from satellite observations: Weather, v. 61, no. 3, p. 70-76, doi:10.1256/wea.222.05

161   Figure from neptune.gsfc.nasa.gov/csb/index.php?section=234

162   Kwok, R., and Untersteiner, N., 2011, The thinning of Arctic sea ice: Physics Today, v. 64, no. 4, p. 36-41, doi:10.1063/1.3580491

163   Derksen, C., and Brown, R., 2012, Spring snow cover extent reductions in the 2008-2012 period exceeding climate model projections: Geophysical Research Letters, v. 39, p. L19504, doi:10.1029/2012GL053387

164   Rignot, E., et al., 2011, Acceleration of the contribution of the Greenland and Antarctic ice sheets to sea level rise: Geophysical Research Letters, v. 38, no. 5, p. L05503, doi:10.1029/2011gl046583

165   Gardner, A. S., et al., 2011, Sharply increased mass loss from glaciers and ice caps in the Canadian Arctic Archipelago: Nature, v. 473, no. 7347, p. 357-360, doi:10.1038/nature10089

166   Fisher, D., et al., 2012, Recent melt rates of Canadian arctic ice caps are the highest in four millennia: Global and Planetary Change, v. 84, p. 3-7, doi:10.1016/j.gloplacha.2011.06.005.

167   Norval, M., et al., 2011, The human health effects of ozone depletion and interactions with climate change: Photochemical & Photobiological Sciences, v. 10, no. 2, p. 199-225, doi:10.1039/c0pp90044c

168   Moore, J. W., Stanitski, C. L., and Jurs, P. C., 2007, Chemistry: The Molecular Science, Volume I, Chapters 1-12, Brooks Cole, 912 p., page 463

169   Figure 7 in McKenzie, R. L., et al., 2011, Ozone depletion and climate change: impacts on UV radiation: Photochemical & Photobiological Sciences, v. 10, no. 2, p. 182-198, doi:10.1039/c0pp90034f

170   McKenzie, R., et al., 1999, Increased summertime UV radiation in New Zealand in response to ozone loss: Science, v. 285, no. 5434, p. 1709, doi:10.1126/science.285.5434.1709

171   Herman, J. R., 2010, Global increase in UV irradiance during the past 30 years (1979–2008) estimated from satellite data: J. Geophys. Res, v. 115, p. D04203, doi:10.1029/2009JD012219

Chapter 8

172   Zielinski, G. A., et al., 1996, A 110,000-year record of explosive volcanism from the GISP2 (Greenland) ice core: Quaternary Research, v. 45, p. 109-118, doi:10.1006/qres.1996.0013

173   White, J. W. C., et al., 1997, The climate signal in the stable isotopes of snow from Summit, Greenland: Results of comparisons with modern climate observations: Journal of Geophysical Research, v. 102, p. 26425-26439, doi:10.1029/97JC00162

174   Mayewski, P. A., et al., 1993, The atmosphere during the Younger Dryas: Science v. 261, p. 195-197, doi:10.1126/science.261.5118.195. Also Mayewski, P. A., et al., 1997, Major features and forcing of high-latitude northern hemisphere atmospheric circulation using a 110,000-year-long glaciochemical series: Journal of Geophysical Research, v. 102, no. C12, p. 26345–26366, doi:10.1029/96JC03365

175   Photo by Dave McGarvie, earthandsolarsystem.files.wordpress.com/2013/02/herdubreid.jpg

176   From figures 3 and 4 in Licciardi, J. M., et al., 2007, Glacial and volcanic history of Icelandic table mountains from cosmogenic ^{3}He exposure ages: Quaternary Science Reviews, v. 26, no. 11-12, p. 1529-1546, doi:10.1016/j.quascirev.2007.02.016

177   Photo by David Harlow, United States Geological Survey.

178   Self, S., et al., 1996, The atmospheric impact of the 1991 Mount Pinatubo eruption, in Newhall, C. G., and Punongbayan, R. S., eds., Fire and Mud: Eruptions and lahars of Mount Pinatubo, Philippines, Philippine Institute of Volcanology and Seismology and University of Washington Press, pubs.usgs.gov/pinatubo/self/, p. 1089-1115

179   McCormick, M. P., and Veiga, R. E., 1992, SAGE II measurements of early Pinatubo aerosols: Geophysical Research Letters, v. 19, no. 2, p. 155–158, doi:10.1029/91GL02790

180   Stone, R. S., et al., 1993, Properties and decay of stratospheric aerosols in the Arctic following the 1991 eruptions of Mount Pinatubo: Geophysical Research Letters, v. 20, p. 2539-2362, doi:10.1029/93GL02684

181   Self, S., et al., 1996, The atmospheric impact of the 1991 Mount Pinatubo eruption, in Newhall, C. G., and Punongbayan, R. S., eds., Fire and Mud: Eruptions and lahars of Mount Pinatubo, Philippines, Philippine Institute of Volcanology and Seismology and University of Washington Press, pubs.usgs.gov/pinatubo/self/, p. 1089-1115

182   Labitzke, K., and McCormick, M., 1992, Stratospheric temperature increases due to Pinatubo aerosols: Geophysical Research Letters, v. 19, no. 2, p. 207-210, doi:10.1029/91GL02940

183   Figure on page 1243 in Robock, A., 2002, Pinatubo eruption: The climatic aftermath: Science, v. 295, no. 5558, p. 1242-1244, doi:10.1126/science.1069903

184   Thompson, D. W. J., and Solomon, S., 2009, Understanding recent stratospheric climate change: Journal of Climate, v. 22, no. 8, p. 1934-1943, doi:10.1175/2008JCLI2482.1

185   Deshler, T., et al., 1993, Balloonborne measurements of Pinatubo aerosol during 1991 and 1992 at 41°N: vertical profiles, size distribution, and volatility: Geophysical Research Letters, v. 20, no. 14, p. 1435-1438, doi:10.1029/93GL01337

186   Based on Figure 1 from Gleckler, P. J., et al., 2006, Krakatoa's signature persists in the ocean: Nature, v. 439, p. 675, doi:10.1038/439675a

187   Based on Figure 5 from Gregory, J. M., et al., 2006, Simulated global-mean sea level changes over the last half-millennium: Journal of Climate, v. 19, no. 18, p. 4576-4591, doi:10.1175/JCLI3881.1

188   Laing, D., 1991, The Earth System: An Introduction to Earth Science, Wm. C. Brown Publishers, 590 p., p. 276-277

189   Lisiecki, L. E., and Raymo, M. E., 2005, A Pliocene-Pleistocene stack of 57 globally distributed benthic δ18O records: Paleoceanography, v. 20, no. 1, doi:10.1029/2004pa001071

190   Lea, D. W., et al., 2000, Climate Impact of Late Quaternary Equatorial Pacific Sea Surface Temperature Variations: Science, v. 289, no. 5485, p. 1719-1724, doi:10.1126/science.289.5485.1719

191   White, J. W. C., et al., 1997, The climate signal in the stable isotopes of snow from Summit, Greenland: Results of comparisons with modern climate observations: Journal of Geophysical Research, v. 102, p. 26425-26439, doi:10.1029/97JC00162

192   Laskar, J., Robutel, P., Joutel, F., Gastineau, M., Correia, A. C. M., and Levrard, B., 2004, A long-term numerical solution for the insolation quantities of the Earth: Astronomy and Astrophysics, v. 428, no. 1, p. 261-285, doi:10.1051/0004-6361:20041335.

193   ftp.ncdc.noaa.gov/pub/data/paleo/icecore/greenland/summit/gisp2/chem/volcano.txt

194   Zielinski, G. A., et al., 1996, A 110,000-year record of explosive volcanism from the GISP2 (Greenland) ice core: Quaternary Research, v. 45, p. 109-118, doi:10.1006/qres.1996.0013

195 Ward, P. L., 2009, Sulfur dioxide initiates global climate change in four ways: Thin Solid Films, v. 517, no. 11, p. 3188-3203, doi:10.1016/j.tsf.2009.01.005

196 Hemming, S. R., 2004, Heinrich events: Massive late Pleistocene detritus layers of the North Atlantic and their global climate imprint: Reviews of Geophysics, v. 42, no. 1, doi:10.1029/2003rg000128

197 Ward, P. L., 2009, Sulfur dioxide initiates global climate change in four ways: Thin Solid Films, v. 517, no. 11, p. 3188-3203, doi:10.1016/j.tsf.2009.01.005

198 Henry, C. D., et al., 1994, 40 Ar/ 39 Ar chronology and volcanology of silicic volcanism in the Davis Mountains, Trans-Pecos Texas: Geological Society of America Bulletin, v. 106, no. 11, p. 1359-1376, doi:10.1130/0016-7606(1994)106<1359:AACAVO>2.3.CO;2

199 mantleplumes.org/Hawaii.html

200 Cogné, J.-P., and Humler, E., 2006, Trends and rhythms in global seafloor generation rate: Geochemistry Geophysics Geosystems, v. 7, no. 3, p. Q03011, doi:10.1029/2005gc001148

201 Veizer, J., et al., 1999, $^{87}Sr/^{86}Sr$, $d^{13}C$ and $d^{18}O$ evolution of Phanerozoic seawater: Chemical Geology, v. 161, p. 59-88, doi:10.1016/S0009-2541(99)00081-9

202 Lea, D. W., et al., 2000, Climate Impact of Late Quaternary Equatorial Pacific Sea Surface Temperature Variations: Science, v. 289, no. 5485, p. 1719-1724, doi:10.1126/science.289.5485.1719

203 Zanazzi, A., et al., 2007, Large temperature drop across the Eocene-Oligocene transition in central North America: Nature, v. 445, no. 7128, p. 639-642, doi:10.1038/nature05551

204 Freda, C., et al., 2005, Sulfur diffusion in basaltic melts: Geochimica et Cosmochimica Acta, v. 69, no. 21, p. 5061-5069, doi:10.1016/j.gca.2005.02.002

205 Ward, P. L., 1971, New interpretation of the geology of Iceland: Geological Society of America Bulletin, v. 82, no. 11, p. 2991-3012, doi:10.1130/0016-7606(1971)82[2991:NIOTGO]2.0.CO;2

206 Hjartarson, Á., 2011, Víðáttumestu hraun Íslands. (The largest lavas of Iceland): Náttúrufræðingurinn, v. 81, p. 37-49

207 Thordarson, T., and Self, S., 2003, Atmospheric and environmental effects of the 1783–1784 Laki eruption: A review and reassessment: Journal of Geophysical Research, v. 108, no. D1, p. 4011, doi:10.1029/2001jd002042

208 Kington, J., 2009, The Weather of the 1780s Over Europe, Cambridge University Press, 180 p.

209   Parker, D., Legg, T., and Folland, C., 1992, A new daily central England
       temperature series, 1772–1991: International Journal of Climatology, v. 12, no. 4,
       p. 317–342, doi:10.1002/joc.3370120402

210   Figure 8 in Thordarson, T., and Self, S., 2003, Atmospheric and environmental
       effects of the 1783–1784 Laki eruption: A review and reassessment: Journal of
       Geophysical Research, v. 108, no. D1, p. 4011, doi:10.1029/2001jd002042

211   Ward, P. L., 2009, Sulfur dioxide initiates global climate change in four ways: Thin
       Solid Films, v. 517, no. 11, p. 3188-3203, doi:10.1016/j.tsf.2009.01.005. Also
       Ward, P. L., 2010, Understanding volcanoes may be the key to controlling global
       warming: Society of Vacuum Coaters Bulletin, v. Summer, p. 26-34

212   Licciardi, J. M., Kurz, M. D., and Curtice, J. M., 2007, Glacial and volcanic
       history of Icelandic table mountains from cosmogenic ^{3}He exposure ages:
       Quaternary Science Reviews, v. 26, no. 11-12, p. 1529-1546, doi:10.1016/j.
       quascirev.2007.02.016

213   Huybers, P., and Langmuir, C., 2009, Feedback between deglaciation, volcanism,
       and atmospheric CO2: Earth and Planetary Science Letters, v. 286, no. 3-4, p.
       479-491, doi:10.1016/j.epsl.2009.07.014

214   Rahmstorf, S., 2003, Timing of abrupt climate change: A precise clock:
       Geophysical Research Letters, v. 30, no. 10, p. 1510, doi:10.1029/2003gl017115

215   Based on Figure 5 in Storey, M., et al., 2007, Timing and duration of volcanism
       in the North Atlantic Igneous Province: Implications for geodynamics and
       links to the Iceland hotspot: Chemical Geology, v. 241, no. 3-4, p. 264-281,
       doi:10.1016/j.chemgeo.2007.01.016

216   Storey, M., et al., 2007, Paleocene-Eocene thermal maximum and the opening
       of the Northeast Atlantic: Science, v. 316, no. 5824, p. 587-589, doi:10.1126/
       science.1135274

217   Röhl, U., et al., 2000, New chronology for the late Paleocene thermal maximum
       and its environmental implications: Geology, v. 28, no. 10, p. 927-930,
       doi:10.1130/0091-7613(2000)28<927:ncftlp>2.0.co;2

218   Sluijs, A., et al., 2006, Subtropical Arctic Ocean temperatures during the
       Palaeocene/Eocene thermal maximum: Nature, v. 441, p. 610-613, doi:10.1038/
       nature04668

219   Bijl, P. K., et al., 2009, Early Palaeogene temperature evolution of the southwest
       Pacific Ocean: Nature, v. 461, no. 7265, p. 776-779, doi:10.1038/nature08399

220   From Figure 1 in Courtillot, V. E., and Renne, P. R., 2003, On the ages of flood basalt events: Comptes Rendus Geoscience, v. 335, no. 1, p. 113-140, doi:10.1016/s1631-0713(03)00006-3

221   Reichow, M. K., et al., 2009, The timing and extent of the eruption of the Siberian Traps large igneous province: Implications for the end-Permian environmental crisis: Earth and Planetary Science Letters, v. 277, no. 1, p. 9-20, doi:10.1016/j.epsl.2008.09.030

222   Joachimski, M. M., et al., 2012, Climate warming in the latest Permian and the Permian–Triassic mass extinction: Geology, v. 40, no. 3, p. 195-198, doi:10.1130/G32707.1

223   Sun, Y., et al, 2012, Lethally hot temperatures during the early triassic greenhouse: Science, v. 338, no. 6105, p. 366-370, page 366, doi:10.1126/science.1224126

224   Svensen, H., et al., 2009, Siberian gas venting and the end-Permian environmental crisis: Earth and Planetary Science Letters, v. 277, no. 3-4, p. 490-500, page 490, doi:10.1016/j.epsl.2008.11.015

225   Beerling, D. J., 2007, The Emerald Planet: How plants changed Earth's history, Oxford; New York, Oxford University Press, xvi, 288 p.

226   Visscher, H., et al., 2004, Environmental mutagenesis during the end-Permian ecological crisis: Proceedings of the National Academy of Sciences of the United States of America, v. 101, no. 35, p. 12,952–12,956, page 12,952, doi:10.1073/pnas.0404472101

227   Reidel, S. P., et al., 2013, The Columbia River flood basalt province: stratigraphy, areal extent, volume, and physical volcanology: Geological Society of America Special Papers, v. 497, p. 1-43, doi:10.1130/2013.2497(1)

228   Newhall, C. G., and Punongbayan, R. S., eds., Fire and mud; Eruptions and lahars of Mount Pinatubo, 1996, Philippines, Philippine Institute of Volcanology and Seismology and University of Washington Press chapters by Self, S., et al., The atmospheric impact of the 1991 Mount Pinatubo eruption, pubs.usgs.gov/pinatubo/self/, p. 1089-1115. Also Scott, W. E., et al., Pyroclastic flows of the 15 June, 1991, paroxysmal eruption, Mount Pinatubo, p. 545–570. Also Gerlach, T. M., et al., 1996, Preeruption vapor in magma of the climactic Mount Pinatubo eruption: Source of the giant stratospheric sulfur dioxide cloud. Also Bureau, H., et al., 2000, Volcanic degassing of bromine and iodine: experimental fluid/melt partitioning data and applications to stratospheric chemistry: Earth and Planetary Science Letters, v. 183, no. 1, p. 51-60, doi:10.1016/S0012-821X(00)00258-2

229   Thordarson, T., and Self, S., 2003, Atmospheric and environmental effects of the 1783–1784 Laki eruption: A review and reassessment: Journal of Geophysical Research, v. 108, no. D1, p. 4011, doi:10.1029/2001jd002042

230   Christiansen, R. L., 2001, The Quaternary and Pliocene Yellowstone Plateau volcanic field of Wyoming, Idaho and Montana: U.S. Geological Survey Professional Paper, v. 729, p. 146 pp. Also Lanphere, M. A., et al., 2002, Revised ages for tuffs of the Yellowstone Plateau volcanic field: Assignment of the Huckleberry Ridge Tuff to a new geomagnetic polarity event: Geological Society of America Bulletin, v. 114, no. 5, p. 559-568, doi:10.1130/0016-7606(2002)114<0559:RAFTOT>2.0.CO;2. Also Jones, M. T., et al., 2006, The climatic impact of supervolcanic ash blankets: Climate Dynamics, v. 27, p. 553-564, doi:10.1007/s00382-007-0248-7

231   Reichow, M. K., et al., 2009, The timing and extent of the eruption of the Siberian Traps large igneous province: Implications for the end-Permian environmental crisis: Earth and Planetary Science Letters, v. 277, no. 1, p. 9-20, doi:10.1016/j.epsl.2008.09.030. Also Black, B. A., et al., 2013, Acid rain and ozone depletion from pulsed Siberian Traps magmatism: Geology, p. G34875. 34871, doi:10.1130/G34875.1. Also Joachimski, M. M., et al., 2012, Climate warming in the latest Permian and the Permian–Triassic mass extinction: Geology, v. 40, no. 3, p. 195-198, doi:10.1130/G32707.1

232   Kump, L. R., et al., 2005, Massive release of hydrogen sulfide to the surface ocean and atmosphere during intervals of oceanic anoxia: Geology, v. 33, no. 5, p. 397-400, doi:10.1130/g21295.1

233   Staehelin, J., et al., 1998, Total ozone series at Arosa (Switzerland): Homogenization and data comparison: Journal of Geophysical Research, v. 103, no. D5, p. 5827-5841, doi:10.1029/97JD02402

234   Douglass, A., et al., 2011, Chapter 2: Stratospheric ozone and surface ultraviolet radiation, in Ennis, C. A., ed., Scientific Assessment of Ozone Depletion: 2010, World Meteorological Organization Global Ozone Research and Monitoring Project - Report No. 52, p. 76

235   Solomon, S., 1999, Stratospheric ozone depletion: A review of concepts and history: Reviews of Geophysics, v. 37, no. 3, p. 275-316, doi:10.1029/1999RG900008

236   Angell, J. K., 1997, Estimated impact of Agung, El Chichón and Pinatubo volcanic eruptions on global and regional total ozone after adjustment

for the QBO: Geophysical Research Letters, v. 24, no. 6, p. 647–650, doi:10.1029/97GL00544

237 Kerr, R. A., 1993, Ozone takes a nose dive after the eruption of Mt. Pinatubo: Science, v. 260, no. 5107, p. 490-491, doi:10.1126/science.260.5107.490

238 Robock, A., 2002, Pinatubo eruption: The climatic aftermath: Science, v. 295, no. 5558, p. 1242-1244, doi:10.1126/science.1069903

239 De Mazière, M., et al., 1998, Quantitative evaluation of the post–Mount Pinatubo NO_2 reduction and recovery, based on 10 years of Fourier transform infrared and UV-visible spectroscopic measurements at Jungfraujoch: Journal of Geophysical Research, v. 103, no. D9, p. 10849-10858, doi:10.1029/97JD03362

240 Johnston, P., et al., 1992, Observations of depleted stratospheric NO_2 following the Pinatubo volcanic eruption: Geophysical Research Letters, v. 19, no. 2, p. 211-213, doi:10.1029/92GL00043

241 Coffey, M. T., 1996, Observations of the impact of volcanic activity on stratospheric chemistry: Journal of Geophysical Research, v. 101, no. D3, p. 6767–6780, doi:10.1029/95JD03763

242 Thompson, D. W. J., and Solomon, S., 2009, Understanding recent stratospheric climate change: Journal of Climate, v. 22, no. 8, p. 1934-1943, doi:10.1175/2008JCLI2482.1. Also Randel, W. J., 2010, Variability and trends in stratospheric temperature and water vapor, in The Stratosphere: Dynamics, Transport and Chemistry: Geophysical Monograph, v. 190, p. 123-135, doi:10.1029/2009GM000870

243 Anderson, J. G., Wilmouth, D. M., Smith, J. B., and Sayres, D. S., 2012, UV dosage levels in summer: Increased risk of ozone loss from convectively injected water vapor: Science, v. 337, p. 835-839, doi:10.1126/science.1222978

244 eoearth.org/view/article/156270/

245 Tabazadeh, A., and Turco, R. P., 1993, Stratospheric chlorine injection by volcanic eruptions: HCl scavenging and implications for ozone: Science, v. 260, p. 1082-1084, doi:10.1126/science.260.5111.1082

246 Freda, C., Baker, D., and Scarlato, P., 2005, Sulfur diffusion in basaltic melts: Geochimica et Cosmochimica Acta, v. 69, no. 21, p. 5061-5069, doi:10.1016/j.gca.2005.02.002

247 Vance, A., et al., 2010, Ozone depletion in tropospheric volcanic plumes: Geophysical Research Letters, v. 37, no. 22, L22802, page 4, doi:10.1029/2010gl044997

248   von Glasow, R., 2010, Atmospheric chemistry in volcanic plumes: Proceedings of the National Academy of Sciences of the United States of America, v. 107, no. 15, p. 6594-6599, doi:10.1073/pnas.0913164107

249   Rasmussen, S. O., et al., 2014, A stratigraphic framework for abrupt climatic changes during the Last Glacial period based on three synchronized Greenland ice-core records: refining and extending the INTIMATE event stratigraphy: Quaternary Science Reviews, v. 106, p. 14-28, doi:10.1016/j.quascirev.2014.09.007 and.iceandclimate.nbi.ku.dk/data/

250   Sánchez Goñi, M. F., et al., 2008, Contrasting impacts of Dansgaard–Oeschger events over a western European latitudinal transect modulated by orbital parameters: Quaternary Science Reviews, v. 27, no. 11-12, p. 1136-1151, doi:10.1016/j.quascirev.2008.03.003

251   iceandclimate.nbi.ku.dk/images/NGD2_tekst_ny1.png

252   Steffensen, J. P., et al., 2008, High-resolution Greenland ice core data show abrupt climate change happens in few years: Science, v. 321, no. 5889, p. 680-684, page 680, doi:10.1126/science.1157707

253   Broecker, W. S., and Denton, G. H., 1989, The role of ocean-atmosphere reorganizations in glacial cycles: Geochimica et Cosmochimica Acta, v. 53, no. 10, p. 2465-2501, doi:10.1016/0016-7037(89)90123-3

254   Levitus, S., et al., 2012, World ocean heat content and thermosteric sea level change (0-2000 m), 1955-2010: Geophysical Research Letters, v. 39, no. 10, L10603, doi:10.1029/2012GL051106

255   Lea, D. W., et al., 2000, Climate Impact of Late Quaternary Equatorial Pacific Sea Surface Temperature Variations: Science, v. 289, no. 5485, p. 1719-1724, doi:10.1126/science.289.5485.1719

256   Environment Canada, 2015, Archive of world ozone maps, exp-studies.tor.ec.gc.ca/clf2/e/ozoneworld.html

257   ozonedepletiontheory.info/ImagePages/eyjafjallajokull-ozone-animation.html

258   Copied from Figure 2b in Sigmundsson, F., et al., 2010, Intrusion triggering of the 2010 Eyjafjallajökull explosive eruption: Nature, v. 468, no. 7322, p. 426-430, doi:10.1038/nature09558

259   De Muer, D., and De Backer, H., 1992, Revision of 20 years of Dobson total ozone data at Uccle (Belgium): Fictitious Dobson total ozone trends induced by sulfur dioxide trends: Journal of Geophysical Research, v. 97, no. D5, p. 5921-5937, doi:10.1029/91JD03164

260   Thorarinsson, S., and Sigvaldason, G., 1972, The Hekla eruption of 1970: Bulletin of Volcanology, v. 36, no. 2, p. 269-288, doi:10.1007/BF02596870

261   Lee, C. T. A., et al., 2009, Constraints on the depths and temperatures of basaltic magma generation on Earth and other terrestrial planets using new thermobarometers for mafic magmas: Earth and Planetary Science Letters, v. 279, no. 1-2, p. 20-33, doi:10.1016/j.epsl.2008.12.020

262   Baragiola, R. A., et al., 2011, Ozone generation by rock fracture: Earthquake early warning? Applied Physics Letters, v. 99, no. 20, p. 204101, doi:10.1063/1.3660763

263   ozonedepletiontheory.info/pre-eruption-ozone.html

Chapter 9

264   en.wikipedia.org/wiki/Dobson_ozone_spectrophotometer

265   atmos.washington.edu/Reed/

266   Reed, R. J., 1950, The role of vertical motion in ozone-weather relationships: Journal of Meteorology, v. 7, p. 263-267, doi:10.1175/1520-0469(1950)007<0263:TROVMI>2.0.CO;2

267   Dobson, G. M., et al., 1929, Measurements of the amount of ozone in the Earth's atmosphere and its relation to other geophysical conditions. Part III: Proceedings of the Royal Society of London. Series A, v. 122, no. 790, p. 456-486, doi:10.1098/rspa.1929.0034

268   Dobson, G., et al., 1946, Bakerian Lecture. Meteorology of the Lower Stratosphere: Proceedings of the Royal Society of London. Series A, Mathematical and Physical Sciences, v. 185, no. 1001, p. 144-175, doi:10.1098/rspa.1946.0010

269   Tonsberg, E., and Langlo-Olsen, K., 1943, Investigations on atmospheric ozone at nordlysobservatoriet Tromsa: Geofys. Publik, v. 13, no. 12, 39 pp.

270   Meetham, A., and Dobson, G., 1937, The correlation of the amount of ozone with other characteristics of the atmosphere: Quarterly Journal of the Royal Meteorological Society, v. 63, no. 271, p. 289-307, doi:10.1002/qj.49706327102

271   Allen, D., and Reck, R., 1997, Daily variations in TOMS total ozone data: Journal of Geophysical Research, v. 102, p. 13-13, doi:10.1029/97JD00632

272   Staehelin, J., et al., 2001, Ozone trends: a review: Reviews of Geophysics and Space Physics, v. 39, no. 2, p. 231-290, doi:10.1029/1999RG000059

273   Shepherd, T. G., 2008, Dynamics, stratospheric ozone, and climate change: Atmosphere-Ocean, v. 46, no. 1, p. 117-138, doi:10.3137/ao.460106

274   Trenberth, K. E., et al., 2007, Observations: Surface and Atmospheric Climate Change, in Solomon, S., et al., eds., Climate Change 2007: The Physical Science Basis. Contribution of Working Group I to the Fourth Assessment Report of the Intergovernmental Panel on Climate Change, Cambridge University Press, p. 235-336.

275   ncdc.noaa.gov/cdo-web/datatools/records

276   Minnis, P., et al., 1993, Radiative climate forcing by the Mount Pinatubo eruption: Science, v. 259, p. 1411-1415, doi:10.1126/science.259.5100.1411

277   Karl, T., et al., 2012, US temperature and drought: Recent anomalies and trends: EOS Transactions of the American Geophysical Union, v. 93, no. 47, p. 473, doi:10.1029/2012EO470001

278   Sallenger Jr, A. H., Doran, K. S., and Howd, P. A., 2012, Hotspot of accelerated sea-level rise on the Atlantic coast of North America: Nature Climate Change, v. 2, p. 884-888, doi:10.1038/NCLIMATE1597.

279   en.wikipedia.org/wiki/2012_Great_Britain_and_Ireland_floods

280   Nghiem, S., et al., 2012, The extreme melt across the Greenland ice sheet in 2012: Geophysical Research Letters, v. 39, no. 20, p. L20502, doi:10.1029/2012GL053611

281   Meese, D. A., et al., 1994, The accumulation record from the GISP2 core as an indicator of climate-change throughout the Holocene: Science, v. 266, no. 5191, p. 1680-1682, doi:10.1126/science.266.5191.1680

282   Chen, D., et al., 2004, Predictability of El Niño over the past 148 years: Nature, v. 428, p. 733-736, doi:10.1038/nature02439

283   Herweijer, C., et al., 2007, North American droughts of the last millennium from a gridded network of tree-ring data: Journal of Climate, v. 20, no. 7, p. 1353-1376, doi:10.1175/JCLI4042.1

284   Brönnimann, S., 2009, Early twentieth-century warming: Nature Geoscience, v. 2, no. 11, p. 735-736, doi:10.1038/ngeo670

285   volcano.si.edu/search_eruption.cfm

286   cru.uea.ac.uk/cru/warming/

287   mbl.is/frettir/innlent/2014/08/19/1_600_earthquakes_in_48_hours/

288   en.wikipedia.org/wiki/Holuhraun

289   Thordarson, T., and Self, S., 2003, Atmospheric and environmental effects of the 1783–1784 Laki eruption: A review and reassessment: Journal of Geophysical Research, v. 108, no. D1, p. 4011, doi:10.1029/2001jd002042

290 Waugh, D. W., and Polvani, L. M., 2010, Stratospheric polar vortices: The Stratosphere: Dynamics, Transport, and Chemistry, Geophysical Monograph Series 190, p. 43-57, doi:10.1029/2009GM000887

291 Figure from blogs.scientificamerican.com/observations/what-is-this-polar-vortex-that-is-freezing-the-us/

292 WOUDC, 2015, World Ozone and Ultraviolet Radiation Data Center, woudc. org/data_e.html. Environment Canada, 2015, Archive of world ozone maps, exp-studies.tor.ec.gc.ca/clf2/e/ozoneworld.html

Chapter 10

293 Climate Change 2013: The Physical Science Basis, ipcc.ch/report/ar5/wg1/, III Glossary

294 Feynman, R. P., et al., 1963, Feynman Lectures on Physics. vol. 1: Mainly mechanics, radiation and heat, Addison-Wesley, page 4-2

295 Ångström, K., 1900, Ueber die Bedeutung des Wasserdampfes und der Kohlensäure bei der Absorption der Erdatmosphäre: Annalen der Physik, v. 308, no. 12, p. 720-732. In English at ozonedepletiontheory.info/Papers/Angstrom1900-English.pdf, doi:10.1002/andp.19003081208

296 Trenberth, K. E., et al., 2009, Earth's global energy budget: Bulletin of the American Meteorological Society, v. 90, no. 3, p. 311-323, doi:10.1175/2008bams2634.1

297 Figure copied from Figure 1 in Trenberth, K. E., and Fasullo, J. T., 2012, Tracking Earth's energy: From El Niño to global warming: Surveys in Geophysics, v. 33, no. 3-4, p. 413-426, doi:10.1007/s10712-011-9150-2

298 Figure copied from Figure 2 of Lüthi, D., et al., 2008, High-resolution carbon dioxide concentration record 650,000–800,000 years before present: Nature, v. 453, no. 7193, p. 379-382, doi:10.1038/nature06949

299 Fischer, H., et al., 1999, Ice core records of atmospheric CO_2 around the last three Glacial terminations: Science, v. 283, no. 5408, p. 1712-1714, doi:10.1126/science.283.5408.1712. Also Monnin, E., et al., 2001, Atmospheric CO_2 concentrations over the last glacial termination: Science, v. 291, no. 5501, p. 112-114, doi:10.1126/science.291.5501.112. Also Caillon, et al., 2003, Timing of atmospheric CO_2 and Antarctic temperature changes across termination III: Science, v. 299, no. 5613, p. 1728-1731, doi:10.1126/science.1078758. Also Siegenthaler, U., et al., 2005, Stable carbon cycle-climate relationship during the Late Pleistocene: Science, v. 310, no. 5752, p. 1313-1317, doi:10.1126/

science.1120130. Also Stott, L., et al., 2007, Southern Hemisphere and deep-sea warming led deglacial atmospheric CO_2 rise and tropical warming: Science, v. 318, no. 5849, p. 435-438, doi:10.1126/science.1143791. Also Tachikawa, K., et al., 2009, Glacial/interglacial sea surface temperature changes in the Southwest Pacific ocean over the past 360 ka: Quaternary Science Reviews, v. 28, no. 13, p. 1160-1170, doi:10.1016/j.quascirev.2008.12.013. Also Pedro, J. B., et al., 2012, Tightened constraints on the time-lag between Antarctic temperature and CO_2 during the last deglaciation: Climate of the Past, v. 8, no. 4, p. 1213-1221, doi:10.5194/cp-8-1213-2012

300   Shakun, J. D., et al., 2012, Global warming preceded by increasing carbon dioxide concentrations during the last deglaciation: Nature, v. 484, no. 7392, p. 49-54, doi:10.1038/nature10915. Also Parrenin, F., et al., 2013, Synchronous change of atmospheric CO_2 and Antarctic temperature during the last deglacial warming: Science, v. 339, no. 6123, p. 1060-1063, doi:10.1126/science.1226368

301   Figure copied from wattsupwiththat.com/2012/04/11/does-co2-correlate-with-temperature-history-a-look-at-multiple-timescales-in-the-context-of-the-shakun-et-al-paper/

302   Zachos, J. C., et al., 2008, An early Cenozoic perspective on greenhouse warming and carbon-cycle dynamics: Nature, v. 451, no. 7176, p. 279-283, doi:10.1038/nature06588

303   Figure based on Figure 5 in Zhang, Y. G., et al., 2013, A 40-million-year history of atmospheric CO_2: Philosophical Transactions of the Royal Society A: Mathematical, Physical and Engineering Sciences, v. 371, no. 2001, doi:10.1098/rsta.2013.0096

304   Knorr, G., et al., 2011, A warm Miocene climate at low atmospheric CO_2 levels: Geophysical Research Letters, v. 38, no. 20, doi:10.1029/2011GL048873

305   Fedorov, A. V., et al., 2013, Patterns and mechanisms of early Pliocene warmth: Nature, v. 496, no. 7443, p. 43-49, doi:10.1038/nature12003

306   Ruddiman, W. F., 2010, A Paleoclimatic Enigma?: Science, v. 328, no. 5980, p. 838-839, doi:10.1126/science.1188292

307   Veizer, J., et al., 1999, $^{87}Sr/^{86}Sr$, $d^{13}C$ and $d^{18}O$ evolution of Phanerozoic seawater: Chemical Geology, v. 161, p. 59-88, doi:10.1016/S0009-2541(99)00081-9

308   en.wikipedia.org/wiki/Sea-level_curve

309   Berner, R. A., 2006, Inclusion of the Weathering of Volcanic Rocks in the GEOCARBSULF Model: American Journal of Science, v. 306, no. 5, p. 295-302, doi:10.2475/05.2006.01

310   Royer, D., 2001, Stomatal density and stomatal index as indicators of paleoatmospheric CO_2 concentration: Review of Palaeobotany and Palynology, v. 114, no. 1-2, p. 1-28, 10.1016/S0034-6667(00)00074-9

311   Crowell, J. C., 1999, Pre-Mesozoic ice ages: Their bearing on understanding the climate system: Geological Society of America Special Paper, v. 192, p. 1-112, doi:10.1130/0-8137-1192-4.1. Also Frakes, L. A., et al., 2005, Climate modes of the phanerozoic: the history of the earth's climate over the past 600 million years, Cambridge University Press, 274 p.

312   Gerlach, T. M., et al., 1996, Preeruption vapor in magma of the climactic Mount Pinatubo eruption: Source of the giant stratospheric sulfur dioxide cloud, in Newhall, C. G., and Punongbayan, R. S., eds., Fire and mud: Eruptions and lahars of Mount Pinatubo, Philippines, Philippine Institute of Volcanology and Seismology and University of Washington Press, p. 415-433

313   PALAEOSENS Project Members, 2012, Making sense of palaeoclimate sensitivity: Nature, v. 491, no. 7426, p. 683-691, doi:10.1038/nature11574

314   Unpublished data compiled by David Laing

315   Rashid, H., et al., 2011, Abrupt climate change revisited: Geophysical Monograph, v. 193, p. 1-14, page 2, doi:10.1029/2011GM001139

316   Serreze, M. C., and Barry, R. G., 2011, Processes and impacts of Arctic amplification: A research synthesis: Global and Planetary Change, v. 77, no. 1, p. 85-96, doi:10.1016/j.gloplacha.2011.03.004

317   Hansen, J., et al., 2005, Earth's energy imbalance: Confirmation and implications: Science, v. 308, no. 5727, p. 1431-1435, doi:10.1126/science.1110252. Also Trenberth, K., and Fasullo, J., 2010, Tracking Earth's Energy: Science, v. 328, no. 5976, p. 316-317, doi:10.1126/science.1187272

318   Kerr, R. A., 2013, Forecasting regional climate change flunks its first test: Science, v. 339, no. 6120, p. 638-638, doi:10.1126/science.339.6120.638

319   Seidel, D. J., et al., 2008, Widening of the tropical belt in a changing climate: Nature Geoscience, v. 1, no. 1, p. 21-24, doi:10.1038/ngeo.2007.38

320   Polvani, L. M., et al., 2011, Stratospheric ozone depletion: The main driver of twentieth-century atmospheric circulation changes in the Southern Hemisphere: Journal of Climate, v. 24, no. 3, p. 795-812, doi:10.1175/2010JCLI3772.1

321   Allen, R. J., et al., 2012, Recent Northern Hemisphere tropical expansion primarily driven by black carbon and tropospheric ozone: Nature, v. 485, no. 7398, p. 350-354, doi:10.1038/nature11097

Chapter 11

322 Michelson, A. A., and Morley, E. W., 1887, On the relative motion of the earth and the luminiferous ether: American journal of science, v. 34, no. 203, p. 333-345, doi:10.2475/ajs.s3-34.203.333

323 Einstein, A., 1905, Über einen die Erzeugung und Verwandlung des Lichtes betreffenden heuristischen Gesichtspunkt: Annalen der Physik, v. 322, no. 6, p. 132-148, doi:10.1002/andp.19053220607. In English at en.wikisource.org/wiki/A_Heuristic_Model_of_the_Creation_and_Transformation_of_Light

324 en.wikipedia.org/wiki/Photoelectric_effect

325 Falkenburg, B., 2010, Particle Metaphysics: a critical account of subatomic reality, Springer, 386 p.

326 en.wikipedia.org/wiki/Planck–Einstein_relation

327 Feynman, R. P., et al., 1963, Feynman Lectures on Physics. vol. 1: Mainly mechanics, radiation and heat, Addison-Wesley, page 4-2

328 Einstein, A., 1905, Über einen die Erzeugung und Verwandlung des Lichtes betreffenden heuristischen Gesichtspunkt: Annalen der Physik, v. 322, no. 6, p. 132-148, doi:10.1002/andp.19053220607. In English at en.wikisource.org/wiki/A_Heuristic_Model_of_the_Creation_and_Transformation_of_Light

329 Evans, J., and Popp, B., 1985, Pictet's experiment: the apparent radiation and reflection of cold: American Journal of Physics, v. 53, p. 737, doi:10.1119/1.14305

330 esa.int/Our_Activities/Space_Science/Planck/Planck_and_the_cosmic_microwave_background

331 Guth, A., 1998, The Inflationary Universe, Basic Books, 384 p. Carroll, S., 2007, Dark Matter, Dark Energy: The Dark Side of the Universe, The Great Courses. Chaisson, E., and McMillan, S., 2011, Astronomy Today, Addison-Wesley

332 Fixsen, D., 2009, The temperature of the cosmic microwave background: Astrophysical Journal, v. 707, no. 2, p. 916, doi:10.1088/0004-637X/707/2/916

333 en.wikipedia.org/wiki/Solar_mass

Chapter 12

334 Evans, J., and Popp, B., 1985, Pictet's experiment: the apparent radiation and reflection of cold: American Journal of Physics, v. 53, p. 737, doi:10.1119/1.14305, p. 738

335 Fourier, J., 1822, The Analytic Theory of Heat. archive.org/details/analyticaltheor00fourgoog

336 Fleming, J. R., 1999, Joseph Fourier, the 'greenhouse effect', and the quest for a universal theory of terrestrial temperatures: Endeavour, v. 23, no. 2, p. 72-75, doi:10.1016/S0160-9327(99)01210-7, page 72

337 en.wikipedia.org/wiki/ File:TyndallsSetupForMeasuringRadiantHeatAbsorptionByGases_ annotated. jpg. Tyndall, J., 1859, On the transmission of heat of different qualities through gases of different kinds: Notices of the Proceedings of the Royal Institution of Great Britain, v. 3, p. 155-158. Tyndall, J., 1861, The Bakerian Lecture: on the absorption and radiation of heat by gases and vapours, and on the physical connexion of radiation, absorption, and conduction: Philosophical Transactions of the Royal Society of London, v. 151, p. 1-36, doi:10.1098/rstl.1861.0001

338 Some examples: Langley, S. P., 1883, The selective absorption of solar energy: American Journal of Science, v. 25, no. 147, p. 169-196, doi:10.2475/ajs. s3-25.147.169. Also Langley, S. P., 1888, LIX. The invisible solar and lunar spectrum: Philosophical Magazine Series 5, v. 26, no. 163, p. 505-520, doi:10.1080/14786448808628306. Also Rubens, H., and Aschkinass, E., 1898, Observations on the absorption and emission of aqueous vapor and carbon dioxide in the infra-red spectrum: Astrophysical Journal, v. 8, p. 176-192, doi:10.1086/140516

339 Arrhenius, S., 1896, On the influence of carbonic acid in the air upon the temperature of the ground: Philosophical Magazine and Journal of Science Series 5, v. 41, no. 251, p. 237-276, doi:10.1080/14786449608620846

340 Crawford, E., 1997, Arrhenius' 1896 model of the greenhouse effect in context: Ambio, v. 26, no. 1, p. 6-11, jstor.org/stable/4314543

341 Ångström, K., 1900, Ueber die Bedeutung des Wasserdampfes und der Kohlensäure bei der Absorption der Erdatmosphäre: Annalen der Physik, v. 308, no. 12, p. 720-732. In English at ozonedepletiontheory.info/Papers/ Angstrom1900-English.pdf., doi:10.1002/andp.19003081208

342 For example: realclimate.org/index.php/archives/2007/06/a-saturated-gassy-argument-part-ii/.

343 Fleming, J. R., 2007, The Callendar Effect: The Life and Work of Guy Stewart Callendar (1898–1964), the Scientist Who Established the Carbon Dioxide Theory of Climate Change, American Meteorological Society, 155 p.

344 Plass, G. N., 1956, Carbon dioxide and the climate: American Scientist, v. 44, no. 3, p. 302-316, jstor.org/stable/27826805

345 Rothman, L., et al., 2013, The HITRAN2012 molecular spectroscopic database: Journal of Quantitative Spectroscopy and Radiative Transfer, v. 130, p. 4-50, doi:10.1016/j.jqsrt.2013.07.002

346 Revelle, R., et al., 1965, Appendix Y4: Atmospheric carbon dioxide, in President's Science Advisory Committee, ed., Restoring the quality of our environment: A report of the Environmental Pollution Panel, p. 111-133. dge.stanford.edu/labs/caldeiralab/Caldeira%20downloads/PSAC, 1965, Restoring the Quality of Our Environment.pdf

347 Maxwell, J. C., 1862, XIV. On physical lines of force: The London, Edinburgh, and Dublin Philosophical Magazine and Journal of Science, v. 23, no. 152, p. 85-95, doi:10.1080/14786446208643219

348 Langley, S. P., 1889, Temperature of the Moon: American Journal of Science, v. 38, no. 228, p. 421-439, doi:10.2475/ajs.s3-38.228.421, p. 424.

349 ipcc.ch/pdf/ar5/prpc_syr/11022014_syr_copenhagen.pdf

350 scienceandpublicpolicy.org/commentaries_essays/crichton_three_speeches.html

351 Montford, A. W., 2010, The Hockey Stick Illusion: Climategate and the Corruption of Science, Stacey Intl, 482 p.

352 Mann, M. E., 2012, The Hockey Stick and the Climate Wars: Dispatches from the Front Lines, Columbia University Press, 448 p. Also Shollenberger, B., 2015, The Hockey Stick and the Climate Wars: Follow-Up on the Hockey Stick, Kindle, 41 p.

353 Figure copied from Figure 1b on page 3 of Folland, C.K., et al., 2001: Observed Climate Variability and Change. In: Climate Change 2001: The Scientific Basis. Contribution of Working Group I to the Third Assessment Report of the Intergovernmental Panel on Climate Change [Houghton, J.T.,et al. (eds.)]. Cambridge University Press, Cambridge, United Kingdom and New York, NY, USA, 881pp, ipcc.ch/ipccreports/tar/wg1/index.htm

Chapter 13

354 Molina, M. J., and Rowland, F. S., 1974, Stratospheric sink for chlorofluoromethanes: Chlorine catalysed destruction of ozone: Nature, v. 249, p. 810-814, doi:10.1038/249810a0

355 Farman, J. C., et al., 1985, Large losses of total O_3 in atmosphere reveal seasonal ClO_x/NO_x interaction: Nature, v. 315, p. 207-210, doi:10.1038/315207a0

356 en.wikipedia.org/wiki/List_of_countries_by_energy_consumption_per_capita

357 knoema.com/smsfgud/world-reserves-of-fossil-fuels

358 flowcharts.llnl.gov

359   who.int/mediacentre/news/releases/2014/air-pollution/en/

360   Lin, M., et al., 2012, Transport of Asian ozone pollution into surface air over the western United States in spring: Journal of Geophysical Research: Atmospheres (1984–2012), v. 117, no. D21, doi:10.1029/2011JD016961. Also Lin, J., et al., 2014, China's international trade and air pollution in the United States: Proceedings of the National Academy of Sciences, v. 111, doi:10.1073/pnas.1312860111

361   Keys, D., 2000, Catastrophe: an investigation into the origins of the modern world, New York, Ballantine Pub., xviii, 343 p.

362   globalwarmingart.com/wiki/File:2000_Year_Temperature_Comparison_png

363   Dull, R. A., et al., 2001, Volcanism, ecology and culture: A reassessment of the Volcán Ilopango TBJ eruption in the southern Maya realm: Latin American Antiquity, v. 12, no. 1, p. 25-44, doi:10.2307/971755. Also Dull, R., et al., 2010, Did the TBJ Ilopango eruption cause the AD 536 event?: American Geophysical Union, Fall Meeting 2010, abstract #V13C-2370, adsabs.harvard.edu/abs/2010AGUFM.V13C2370D

364   en.wikipedia.org/wiki/Eldgjá

365   en.wikipedia.org/wiki/2010_eruptions_of_Eyjafjallajökull

366   en.wikipedia.org/wiki/Toba_catastrophe_theory. Also volcanoes.usgs.gov/images/pglossary/eruptionsize.php

367   Dawkins, R., 2004, The ancestor's tale: a pilgrimage to the dawn of evolution, Boston, Houghton Mifflin, xii, 673 p.

368   volcanoes.usgs.gov/volcanoes/yellowstone/yellowstone_geo_hist_50.html

369   en.wikipedia.org/wiki/List_of_largest_volcanic_eruptions

370   Job 12:8 carved in stone over the entrance to Schermerhorn Hall at Columbia University, home to the Department of Geology since 1897, now the Department of Earth & Environmental Sciences. Image from Bwog, Columbia Student News, images.bwog.com/wp-content/uploads/2014/03/IMG_2846.jpg.

INDEX

Note: Page numbers in *italics* refer to illustrative matter.

The Geological Society of America 2012 Geologic Time Scale
From 0 to 252 million years ago

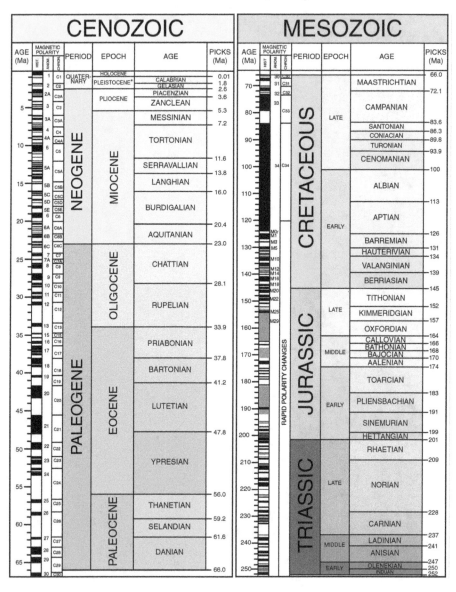

The Geological Society of America 2012 Geologic Time Scale
From 252 to 4100 million years ago

PALEOZOIC

AGE (Ma)	PERIOD	EPOCH	AGE	PICKS (Ma)
	PERMIAN	Lopingian	CHANGHSINGIAN	252
			WUCHIAPINGIAN	254
260		Guadalupian	CAPITANIAN	260
			WORDIAN	265
			ROADIAN	269
				272
280		Cisuralian	KUNGURIAN	279
			ARTINSKIAN	
				290
			SAKMARIAN	296
300			ASSELIAN	299
	CARBONIFEROUS PENNSYL-VANIAN	LATE	GZHELIAN	304
			KASIMOVIAN	307
		MIDDLE	MOSCOVIAN	
320				315
		EARLY	BASHKIRIAN	323
	CARBONIFEROUS MISSIS-SIPPIAN	LATE	SERPUKHOVIAN	331
340		MIDDLE	VISEAN	
				347
		EARLY	TOURNAISIAN	359
360	DEVONIAN	LATE	FAMENNIAN	
				372
380			FRASNIAN	383
		MIDDLE	GIVETIAN	388
			EIFELIAN	393
400		EARLY	EMSIAN	408
			PRAGIAN	411
			LOCHKOVIAN	419
420	SILURIAN	PRIDOLI		423
		LUDLOW	LUDFORDIAN	426
			GORSTIAN	427
		WENLOCK	HOMERIAN	430
			SHEINWOODIAN	433
		LLANDO-VERY	TELYCHIAN	439
440			AERONIAN	441
			RHUDDANIAN	444
			HIRNANTIAN	445
	ORDOVICIAN	LATE	KATIAN	453
460			SANDBIAN	458
		MIDDLE	DARRIWILIAN	467
			DAPINGIAN	470
		EARLY	FLOIAN	478
480			TREMADOCIAN	485
	CAMBRIAN	FURON-GIAN	AGE 10	490
			JIANGSHANIAN	494
			PAIBIAN	497
500		Epoch 3	GUZHANGIAN	501
			DRUMIAN	505
			AGE 5	509
		Epoch 2	AGE 4	514
520			AGE 3	521
		TERRE-NEUVIAN	AGE 2	529
540			FORTUNIAN	541

PRECAMBRIAN

AGE (Ma)	EON	ERA	PERIOD	BDY. AGES (Ma)
	PROTEROZOIC	NEOPRO-TEROZOIC	EDIACARAN	541
				635
750			CRYOGENIAN	
				850
			TONIAN	
1000				1000
		MESOPRO-TEROZOIC	STENIAN	
				1200
1250			ECTASIAN	
				1400
1500			CALYMMIAN	
				1600
		PALEOPRO-TEROZOIC	STATHERIAN	
1750				1800
			OROSIRIAN	
2000				2050
			RHYACIAN	
2250				2300
			SIDERIAN	
2500	ARCHEAN	NEOARCHEAN		2500
2750				2800
3000		MESO-ARCHEAN		
3250				3200
		PALEO-ARCHEAN		
3500				3600
3750		EOARCHEAN		
4000	HADEAN			4000

Printed in the USA
CPSIA information can be obtained
at www.ICGtesting.com
JSHW011303290724
67221JS00010B/389